Prize-winning Science Fair Projects

Prize-winning Science Fair Projects

Penny Raife Durant

SCHOLASTIC INC.
New York Toronto London Auckland Sydney

ISBN 0-590-44019-5

12 11 4 5 6/9

Printed in the U.S.A. 40

First Scholastic printing, November 1991

For my son, Geoffrey, who introduced me to the joy of science and science fairs, with love.

Table of Contents

Acknowledgments

The author would like to thank her husband, Omar, for his help with the illustrations. A great debt of gratitude is owed to Paul Mitschler, consultant to the project and wonderful science teacher, for his invaluable expertise and guidance. Thanks go also to Patricia Raife, Randi Buck, Amy Larson, and Paul Melendres for their contributions.

Introduction

This book was written to help you do your science fair project. If this is your first time doing one, it will get you off to a good start. It explains all the parts of the project, from choosing an idea to presenting your work. If you've done science fair projects before, it will help you improve your project and your chances of winning a prize. Even if you don't win a prize, you'll learn a lot about how scientists and researchers work. A science fair project can be a lot of work, but it can be a lot of fun as well. Choosing the project that's right for you will help you enjoy your work.

Use this book as a guide. Add your own creativity and knowledge. Read and reread the pages about safety. Some areas of science can be dangerous if you don't take safety precautions. Never work alone.

Most of all, enjoy yourself. Science is fascinating. If you put your creativity to work, you may find out something important that no one has known before. That's what scientific research and science fairs are all about.

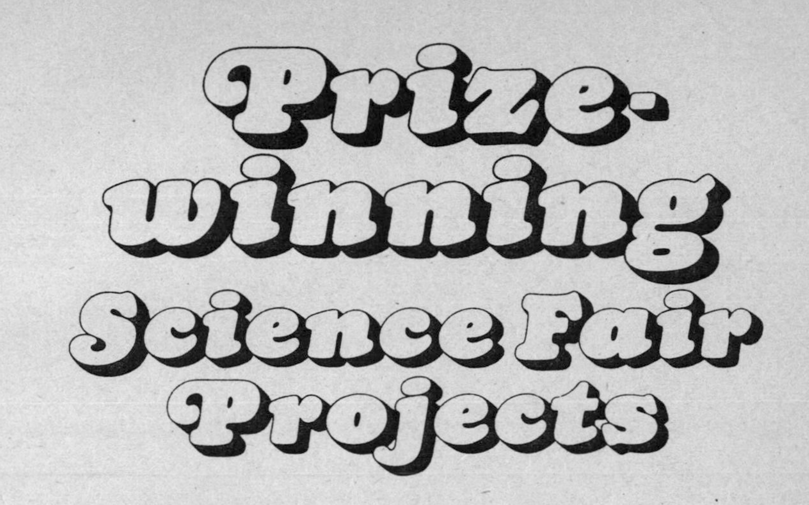
Prize-
winning
Science Fair
Projects

1.
What Is a Science Fair Project?

The dates for your school's annual science fair have been set. Your teacher requires everyone in your class to do a science fair project. You have several months to work on it, and you can choose your own topic. You're interested, but not quite sure what you'll have to do. Just what *is* a science fair project?

A science fair project asks a question. "What Does Yeast Need to Grow?" "Can Earthworms Learn?" "How Soluble Is Your Calcium Supplement?" are all questions that science fair projects could answer. Your science fair project should ask a question that interests you. It should be specific and something that *can* be answered.

Let's say you're interested in plants. "Plants" is too broad a topic for your project. Narrow your focus. You want to grow plants. Okay. That's better, but which plants? "Growing Marigolds" is a project about plants.

Let's narrow the focus even more. "Do marigolds develop greater root systems when watered with plain water or with a plant food solution?" asks a very specific question. You've narrowed the subject of plants to a specific type of plant. You've narrowed the subject of growing plants to root development. This might be a project you'd like to do.

A science fair project also answers the question asked. The answer reflects research and experimentation designed to answer the question. If you ask, "How does exercise affect my pulse rate?" you must limit your answer to just that.

The best science fair projects are one of two things: a research project or an experiment. Most judges prefer experiments, but research projects can be creative as well. Some science fairs allow models, collections, or demonstrations. Some don't. If yours does, you should know it can be difficult to make them original and interesting. Usually judges don't rate them high on creative ability.

A research project means just that. You'll answer your question by doing research. You won't perform an experiment and gain new information. Instead, you'll put together available information in a new way. An example might be, "How much caffeine do I get when I drink a can of soda pop?" To do this experiment, you would have to research the amount of caffeine in soda pop. You would have to be clever and persistent in getting information. You might write to the beverage companies or dig out information in some other way.

An experiment project asks a question and answers it after the researcher has performed some sort of test. Most of the projects in this book are this type. Experiments usually show the most original thinking. All science fairs accept and encourage experiments.

Any of these types of projects will help you to understand the work of a scientist. Scientists are sometimes referred to as detectives. Like detectives, they try to get to the bottom of things. They further our understanding of the world, answer questions, and make discoveries.

When you do a science project, you learn more about a particular area of science. You may develop an interest in a new area of science or discover something important. And, you will experience the thrill of detective work — searching and finding the answer to a puzzling question.

If you are interested in working with people, animals, chemicals, DNA, or bacteria, you will need to check with your teacher about the official International Science and Engineering Fair rules and approval you must get before you present your work in a regional or state science fair.

2.
The Scientific Method

When you do your science fair project, you will use a system of investigation that many scientists use. It is called the scientific method. The scientific method is an outline to help you plan. It is a process for getting data. And it provides a structure for presenting your findings.

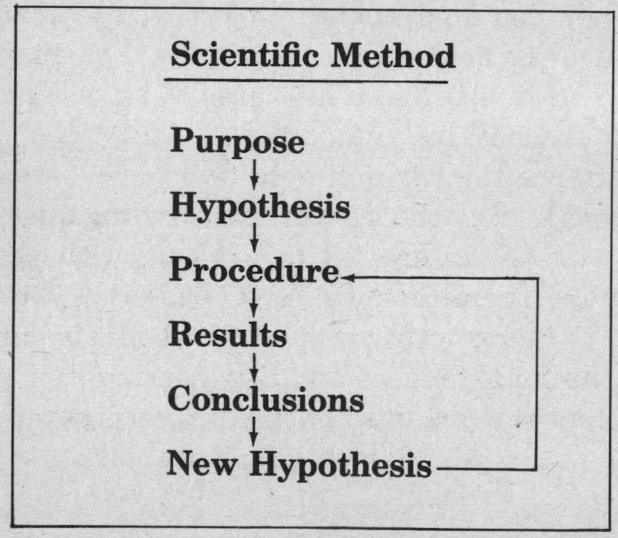

Scientific Method

Purpose
↓
Hypothesis
↓
Procedure ←
↓
Results
↓
Conclusions
↓
New Hypothesis → (back to Procedure)

There are four basic steps to the scientific method. The first is to develop a *purpose and hypothesis.* Once you ask your question, you will know your purpose. The purpose of "Do marigolds develop greater root systems when watered with plain water or with a plant food solution?" is to determine whether plant food helps marigolds to grow.

Your hypothesis is a possible solution, or your educated *guess* at the answer. One hypothesis for the marigold project might be, "Marigolds will develop greater root systems when watered with a plant food solution." Remember, at this point your hypothesis is a guess. It is something that must be tested.

The *procedure* is the next step in the scientific method. You will do research and design an experiment to test your hypothesis. Research means reading books and articles about your subject. For the marigold project, you will read about marigolds, watering plants, and plant food.

You may find that another scientist has done the same project. That's okay. You will be able to compare your results with those of the other scientist. This often occurs in the world of science. One scientist's work may help prove another's to be right, or it may raise questions about the work. In that case, both experiments would need to be repeated.

Designing an experiment may sound hard. This book will help you by showing you step-by-step how to do some experiments.

The third step in the scientific method is to collect and record the *results* of your experiment. Do your experiment and keep track of what you did and what happened. A project notebook will help you when it comes time to put the project together into an exhibit.

For the marigold project, you'll keep track of the amount of water and plant food you give your marigolds by writing it down. You'll write down other things about the plants, too. Include observations such as the formation of leaves or buds, even though they won't be included in your results. Results usually deal with numbers and measurements, and can be presented in graphs or charts or photographs. Scientists use *metric* measurements that are accepted worldwide. You'll want to use metrics, too. If you grow marigolds, you'll weigh the roots of each of the plants *in grams*. You can chart the weights.

The fourth step in the scientific method is to draw *conclusions* based on your results. You will see if one set of marigolds weighs significantly more than the others. This will lead you to accept or reject your hypothesis. Rejecting your hypothesis is okay. It doesn't mean you have a bad project.

Many award-winning projects reject the initial hypotheses. Based on your understanding at this point, you'll try to explain why the experiment supported or rejected your hypothesis. What did your project show, and what might this mean for someone else interested in the subject?

Don't be frightened by the term *scientific method*. It is a process to help you in your work. By proceeding in an orderly, scientific fashion, you will make the job easier on yourself. Also, your results are more likely to be accurate.

3.
Choosing a Topic

When you think about a topic for your project, remember that you'll be working on it for a long time. Don't choose something that won't hold your interest.

The first step is to identify an area that interests you. What are you interested in? What is your background? If you're interested in livestock, do you live on a farm? It might be foolish for a city student to do an experiment involving cattle. What is available to you? Try to narrow down your area of interest, then brainstorm ideas with a friend, your parents, or your teacher.

The next step is to go to the library. Look for books and articles about your topic. Ask the librarian for help in finding information. The more specific you are, the more helpful the librarian can be.

At this point you must ask yourself: Is there enough information available about this topic to make it worthwhile? You don't need to know a great deal about something before you start. You can learn about something new to you. By doing research, you can learn a great deal about your topic. The more you find out about a subject, the deeper and more specific your project will be. If you don't do much research, your project may be too broad and uninteresting.

Find an adult who will be willing to work with you. Talk over your ideas with the adult. You'll do the research and experiment, but your adult partner will be with you all the way. **Science Safety Rule Number One is: NEVER WORK ALONE.** It is important to note, however, that the International Science and Engineering Fair rules do not allow two students to enter work together. Check with your teacher if you're in elementary school. You may be able to work with a friend on a project if you enter an elementary school science fair. If you're older, you'll have to do your own project. But you'll still want an adult partner for safety's sake.

You can get support from researchers and professionals in the field of your topic. Consult with them. If you've done your research and narrowed your topic, a professional will probably be happy to help you. A professional can give you advice or lead you to more information. He or she can lend you specific equipment or supplies. But the professional will not do the experiment for you.

Ask yourself: Are the necessary specimens, equipment, chemicals, etc., available to carry out this project?

Also ask yourself if your project is relevant. We depend on scientists to help us find better ways of doing things. We need more efficient ways to use our resources. New technology can improve our lives and help us to live longer. Where does your idea fit in the real world? Talk this over with your adult partner before you commit yourself to a project.

For example, if you live in the desert, you know that a lot of water is used to make grass and flowers grow. Could you test different watering systems to determine which is most efficient in using the precious resources of water? Could you find out about plants that need little water versus plants that need a lot? Could you design an experiment to test them?

4.
Research

As mentioned before, the quality of your project will depend on the quality of your research. The more you learn about your topic and the more you can tell others about it, the better it will be.

Before you start looking for information, you need to be ready to record that information so you can use it. You need to get organized. First, you'll need a *project notebook*. A loose-leaf binder, with separate sections for different parts of your work, is best. Have plenty of paper, sharp pencils, an eraser, and a package of index cards. Keep your things together and always have them handy. A wonderful piece of information can easily be lost if it's written on a scrap of paper.

So where do you begin? Research, for most of us, begins in a library. Begin with your school library, which is probably familiar to you. The school librarian knows what students need to know in order to find things.

Don't rely on just your school library, though. Look in your phone book under libraries. Look in the section for government agencies and then under schools, colleges, and library services. In all these places you may find libraries or schools with libraries.

If you've focused your thinking on a particular area, you'll know what kind of specialized library will help you most. If you're doing a project that deals with medicine, look for the closest medical library. If you live too far from such a library, don't worry. Most of the information you'll need will be available in public libraries and through library exchange programs.

Before you go to the library, make a list of all the possible subject names you might look up. For example, if your project is about plant growth in different light, think about all the words that describe plant growth, such as: plant growth, botany, plants, light, houseplants, and agriculture. Make sure this list of *key words* reflects what you are studying.

Then go to the card catalog. Many libraries have switched to microfiche or computer terminals instead of card catalogs. If you don't know how to use these, ask a librarian. Also check the *Readers' Guide to Periodical Literature*. Write down the call numbers of books and their titles in your notebook. Gather as many titles as you can. Then go to the shelves to see if they are in the library. Take the books to a table and check through them to see if they are going to be helpful. You can't always tell by the title. Check the Table of Contents and the Index for your key words.

If you're new to a subject, begin with the simplest books that cover broad topics. For the plant growth project you might find a book in the children's section titled, "Plants." By reading it, you may learn more about how plants grow. Then you'll be able to understand the more specific things with which you'll be experimenting. For a lot of quick information about a general topic, use the encyclopedia. Look for key words at the end of the entries you read. Follow up on them and do more reading. Add these key words to your list if they apply to your topic.

Other sources of information include newspaper and magazine sections about science and medicine; educational television; science books; and free government pamphlets.

After you've read generally about your topic, try to locate books that relate to your area of *focus*. You should be able to understand them better having read books that are less specific. Probably the most specific information on research being done in your area will be found in scientific journals. You will find these in libraries for professionals such as university libraries or hospital libraries.

Think about where the professionals in this field would go for information. Ask around.

The older you are, the more will be expected of you. Once you enter middle school or junior high school, you will be expected to read adult books and articles in journals for professionals. Don't worry if you don't understand it all. Take notes. Write questions in your project notebook. The adult who helps may be able to explain some of the more difficult material.

Be sure to keep a record of every book or article you read. Use three- by five-inch cards to record the information for the Bibliography. Use one card for each reference book or article. Keep a rubber band around the cards and put them in a box. You don't want to do this work twice!

When you talk with a professional, write down the person's name, position or title, company or university, phone number, and the date you called.

Use other cards to take notes about your topic. Write one fact per card and put the author's name on the top of the card. When you go to organize your research paper, you'll be able to move these cards around to suit your outline. It will make writing the report much easier.

While you are making notes about your subject, keep a project notebook as well. This notebook will be a record of what you've done. Use it to help you. Write down notes of interest. As yourself questions or write down questions you will want to ask a professional. Once you begin your experiment, use the notebook to make drawings or sketches of what is happening, notes on progress, data, your observations, guesses as to what is happening, questions about why things are happening the way they are, materials you're using, expenses you've had, etc. Tuck your photographs into a pocket folder in this notebook. Keeping everything together will make your work easier.

Now, after you've done some research, you need to find a professional to answer some things you just can't do alone. What if your adult helper doesn't know either? Where do you look? Begin at school with your science teacher. Then ask yourself who would be likely to know about your field. Remember those key words? They should help you in your search. Think about the businesses in your community that use the skills you're studying. For instance, plant growth is important to nurseries, greenhouses, and botanists. Check the yellow pages of your phone book. Call the university or college near you. Sometimes a student at a local college will be able to help out.

First, be sure you've done your homework so you can ask intelligent questions. A person who works in a nursery or greenhouse can suggest types of houseplants that would be good to use if you want quick growth of leaves or large root systems.

Know what you want to do. Tell the professional your idea. Then listen to his or her suggestions and take notes.

Know what expenses may be involved. Now is the time to refocus your project if you cannot afford to buy the materials involved. Some things can be borrowed or made. Many things are available in your home or from your school. Check on this before you commit yourself to a certain project. This is all part of your research.

Remember that research will be a part of your project up until the end. You will want to continue reading about your area even after you've designed your experiment. You may even find a way of changing the experiment for another year's project.

5.
A Safety Check Sheet

READ THIS CAREFULLY
BEFORE YOU BEGIN ANY PROJECT.

Science fair projects are a lot of fun. You'll enjoy working on your project. But before you begin, you need to think about safety. Many areas of science are dangerous as well as fascinating. In all areas, you must be careful. Think through your experiment and what it will involve before you start. Talk about safety precautions with your adult partner. Read the entire list below. Note the safety rules that will apply to your project.

1. Never work alone.

2. Check your equipment carefully. Look for sharp edges, wiring that is frayed or cracked, broken glass, or other hazards.

3. Make sure all electrical equipment is functioning properly before you begin.

4. Don't work alone with chemicals or acids. Be sure you have checked with a chemist or science teacher before you mix any chemicals. Seemingly harmless chemicals can become volatile when combined.

5. Don't keep extra chemicals around. Get only what you need and return any you don't use. Be especially careful around younger children and pets.

6. Is it possible your project will splatter or have a chemical reaction? Use goggles to protect your eyes. If you're working with concentrated solutions, caustic materials, or substances that can stain, wear rubber household gloves and a smock or apron.

7. Find out and follow proper disposal techniques for chemicals, agar plates, and solutions. Most should NOT be put out with the garbage or flushed down the drain.

8. Get approval from the region or state science fair before you begin any project involving people or animals, chemicals, DNA, or bacteria.

9. Design your project to treat animals or humans in a humane way.

10. Do not include any dangerous items in your display. Instead, include photographs of things you used.

11. You will need expert help if you plan to do a blood-related project. You will also have to have authorization from science fair officials.

12. Use a technical laboratory — not your home — for any part of your experiment that could possibly be dangerous, such as mixing chemicals.

13. Clean up after yourself. Don't leave parts of your project lying around for others to hurt themselves or to ruin your work. If you have a younger brother or sister, or a pet, you'll need to find a safe place to keep your experiment and your project.

14. The more dangerous your project, the more help you should seek. It is better to have your adult with you, watching, than to try things alone and get hurt.

15. If you have a question in your mind about the safety of a procedure or materials, wait until you have checked it out with an adult.

BE SAFE, NOT SORRY!

6.
The Experiment

You've decided on the topic you want to study. You've done your research, so you know how you want to focus your work. You have an adult to work with you. It's time to design your experiment.

First, you must break your idea into smaller, manageable parts. For example, if you want to study the effects of plant food on the growth of plants, you will have to narrow your subject to a particular plant, say a pothos, which is an inexpensive and hardy houseplant. Now, before you begin designing your experiment, you'll have to decide what you want the focus to be. Do you want to measure the number of new leaves that appear on the plant? Do you want to weigh the root system? Do you want to measure the total weight of the plant?

Each of these suggestions would be possible. You'll need to choose only one. Remember that you want to be specific.

Let's say you decide to count the number of new leaves that develop. You've narrowed your topic and given a focus to your experiment. From your research you know what a plant needs to thrive. You've made notes about plant growth and learned about pothos in particular.

One of the first things you must now decide is how much time you're going to allow for the experiment. When measuring growth, you'll have to allow a long enough period of time to be sure you'll have enough data. You might have to allow several months.

You'll need to know what the variables might be in this experiment. Knowing what a plant must have in order to grow, you can list several variables. Even though you're focusing on the effect of plant food, you should list all the other variables as well. Here's a partial list of variables that produce different plant growth:

- amount, intensity, and color of light
- quality of water
- type and age of plant
- temperature of environment and soil
- watering method, pattern, or frequency
- type and amount of fertilizer
- quality of soil

Because you want to focus on plant food, which

is fertilizer, keep every other variable as uniform as possible. You should keep all your plants in the same place so they receive the same light and temperature. Be sure the soil is the same for all your plants. Decide how much to water them, and give them all the same amount. Rotate the plants in the location you've chosen to keep the light and heat as equal as possible. The only difference then should be the addition of plant food.

USING A CONTROL

In order to have data that you can use in making judgments about what happened, you will need to use a *control*. For instance, in the plant experiment you would have at least two plants. One of them you would water with ordinary tap water, and one with a plant food solution. The plant that gets only water is your control. It will show what would have happened if you did not experiment with plant food.

Scientists often use a control. For instance, in testing pharmaceuticals, doctors will give placebos to one group of volunteers in a study. Placebos are dummy pills that should have no effect at all. The other group will be given the medicine that the researchers want to study. The difference between the groups is then determined.

You will determine the difference in number of new leaves between your two groups of plants.

HYPOTHESIS

Before you design your experiment, you will need to develop your hypothesis. Based on your research, what do you *think* will happen? When you make your guess, use words that tell the judges and observers what you will measure. If you want to measure the number of leaves, you might make the following hypothesis:

> The pothos plant that is fed with a plant food solution according to the package instructions will develop significantly more new leaves than the pothos plant watered with ordinary tap water.

You are telling the observer what you measured (the number of new leaves) and what you used as a variable (plant food solution).

TESTING YOUR HYPOTHESIS

With the help of your adult partner, set up a convenient place for your experiment. Plan a space that will not get in the way of daily living in your household. Reread the safety check sheet. Plan exactly what you are going to do. Write it all down so that you won't forget. Mark your calendar for special days, such as the days you will water the plants and the day you will end your experiment.

Be sure to follow your plan. Measure your materials, the water, and plant food, every time. Faithful attention to details can make a big difference in the quality of your experiment.

RESULTS

The results section of your experiment simply tells what happened. Results deal in numbers, which you will measure or count. You'll use metric measurements just as scientists do. You'll report your results by answering questions such as: How many? What is the difference? How long? For the project we've been discussing, you would ask: How many new leaves developed? Your results would show numbers that you could then compare.

It is important to keep notes about the growth of the plants. You may find that the same number of leaves grew on both plants, but that the leaves on the fertilized pothos were larger. Or you may find that fewer old leaves died on the tap water plant. These are interesting things to note in your notebook. You might want to test these findings next time.

However, you won't report those happenings in your Results section. You're focusing your project on the number of new leaves, and that's what you must report.

Keep close track of these numbers. Mark them down in your project notebook. You will need them when you get ready for your presentation.

CONCLUSIONS AND INTERPRETATIONS

This is the section of your experiment where you answer the question that your hypothesis proposed. You either prove or disprove your hypothesis. Did the addition of plant food solution make a difference?

Based on your research, you will discuss why you think you got the answer you did. Here, after proving or disproving your hypothesis, you can discuss other aspects of the experiment. Perhaps you obtained the same number of leaves, but the fertilized plant had larger, darker leaves. This is the place to mention that, and to state why you think it happened.

We'll talk more about Conclusions in the next chapter, which deals with the Report and Presentation.

7.
The Report

The first part of the presentation of your work is your report. The second part is a physical display. The third part involves talking with the judges.

Your experiment is complete. You've measured the difference between your control group and your experimental group. You have an idea from your research about why you got the results you did. Now it's time to think about that report.

The report is important for two reasons. One is that it will help you organize your thoughts and data. In writing your report, you'll find out if you indeed understood your topic and your experiment. The other reason is that scientific knowledge needs to be shared. If you hadn't been able to read all the articles and books you read about your subject, you would have had to repeat everyone's experiments and observations for yourself. Think how wasteful that would be! By writing up what you've done, you're performing an important part of scientific work. A scientific journal probably won't want to publish your work yet, but you're gaining skills that will serve you later if you decide to be a scientist.

THE ABSTRACT

The first page of your report is the *Abstract.* Actually, it will probably be the last thing you write. After you've written up all the other sections, you'll have a better idea of what you want to say in the abstract.

An Abstract is a brief description of what you did and what you found out. Most science fairs set a limit of 250 words. It will include your hypothesis, a short paragraph about your experiment, your results, and your conclusion. You can see why you would want to wait to write it.

TITLE PAGE

Next comes your Title Page. Do not put your name on it. Many science fairs require that no name appear anywhere on the exhibit. Your entry form will tell the officials which project is yours.

Make your title intriguing. Many people suggest you ask a question in your title, as has been done in the projects in this book. Or you may want to word it in a different way. Another beginning for a title is "The Effects of . . ."

If you are in middle school or junior high, you'll have to title your projects differently from what you did in elementary school. You'll need to be more specific and use scientific terms. You'll find that your purpose and your title will be quite similar. For example, the title, "Can Earthworms Learn?" might be retitled using the scientific name for earthworms. The title, "Are a Doctor's Hands Really Clean?" might become, "Which Antibacterial Soaps Destroy the Most Surface Bacteria on Skin?"

Just remember that the title will catch the eye of the observers and judges. It's the first thing they notice, so give it a lot of thought.

TABLE OF CONTENTS

The Table of Contents page will come next. Again, you won't be able to write it until you've finished writing the rest of the paper. The Table of Contents helps the judges refer to a specific section of your work. Maybe they will want to read more about your results or about your conclusion and interpretations. Help them by making your Table of Contents accurate and neat.

ACKNOWLEDGMENTS

It's important to give credit to those who assisted you, including your adult partner and any others who helped in your research or experiment. This is a page that formally thanks them and lets the judges know that you made use of the skills and expertise of others.

A good way to word your thanks would be something such as: "I gratefully acknowledge the help of Mr. John Smith, who helped me design and build the maze for my experiment."

EXPERIMENT REPORT

The experiment report tells everything you did, in detail. It is an organized presentation of your work. It includes the Purpose, Materials, Procedure, Results, Conclusions, Discussion, and Bibliography.

Purpose. A statement of the Purpose of your experiment comes after the Acknowledgments. State in a very simple way just what you intended to do. Look at the experiments in this book for examples.

Materials. Make a list of all the Materials you used. Make a diagram or take photographs of any equipment you made specifically for the experiment. The purpose of this section is to make it possible for another researcher to duplicate your experiment. In the scientific world, experiments have to be repeatable before they are accepted as fact. You may have heard about the controversy over some scientists' claiming to have produced cold fusion. Because others could not duplicate the scientists' results, they were skeptical.

Procedure. Outline what you did in your experiment. Be sure to mention how you controlled all the variables except the one you were testing. This is especially important if you reject your hypothesis. It will show that you were careful in your work. For instance, in the plant food versus tap water experiment, mention that the plants were placed in the same place to receive the same amount of light, and were watered on a regular schedule.

Results. This section shows the data you collected. It should include words that describe your findings as well as charts and graphs you've made to show the difference between your control and experimental groups.

Don't interpret or explain the information here; just report it. Make your presentation as clear as possible.

Remember that results deal with numbers or measurements. They alone don't answer the question you originally asked. You'll need to check these numbers so you can answer some other questions, such as: "What do the results mean?" "Are the differences between my control and my experimental groups signficant?"

Scientists use many different methods of testing their results. Talk with your math teacher about statistics and analyzing your results in a mathematical way. You'll find out if the difference between your groups is significant. Running a fairly simple test on your numbers will take your project one step further toward being professional.

Conclusion and Discussion. Now you can answer your original question. Based on the data you collected, did you prove or disprove your hypothesis? Explain why you think you got the answer you did. Mention any problems that occurred that might have affected your results. Refer back to your research when you discuss your findings. Remember that your project grew from your research. This is the place to tie it all together.

It would be a good idea to show your results to a professional and discuss what happened and why.

This section allows you to tell the judges and others what you feel about the experiment. How would you do it differently next time? How might you control it better? How could you extend it? What does the information you gathered mean to people? Can they use your findings in some practical way? The Discussion lets you tell all about these things.

Bibliography. Ask your teacher for a standard form for your Bibliography. In it you will include all the references from your note cards for the books and articles you read. It will be in alphabetical order by the authors' last names.

When others want to review what you did, they can use your Bibliography to find your references. A Bibliography not only acknowledges the work of others, it helps future researchers to find original information.

RESEARCH REPORT

A research report may not be required by your school science fair. (An experiment report will always be needed.) Check with your teacher about the research report. If you're in middle school or junior high school, you will want to do one.

The research project is where you write up what you learned before you began to experiment. Use your note cards to help you. Refer to the work that others have done. Give them credit in the Bibliography and by name in the body of your work.

Begin by making an outline. Fit your note cards into the outline, and then begin writing, filling the spaces in your outline with information you have learned.

After you have written about your subject, include any graphs, charts, pamphlets, or tables of information you copied from reference materials.

8.
Display

According to the rules of the International Science and Engineering Fairs, your display should stand by itself. The outside dimensions must not exceed 48 inches (122 cm) wide, 30 inches (76 cm) deep, and 108 inches (274 cm) high. If your display will sit on a table, you must count in about 30 inches (76 cm) for the table height. You may want to check the rules of your local fair. Remember, you must transport the display to the science fair. Don't make it so large that you can't get it there.

BACKDROP

Your backdrop is the backbone of your display. On it you will put everything else. You may make your backdrop as simple or as fancy as you wish. The simplest backdrop is made by cutting the top, bottom, and one side off a large cardboard carton. Be sure you measure it first. If you are in junior high or middle school, you will want to do more than this.

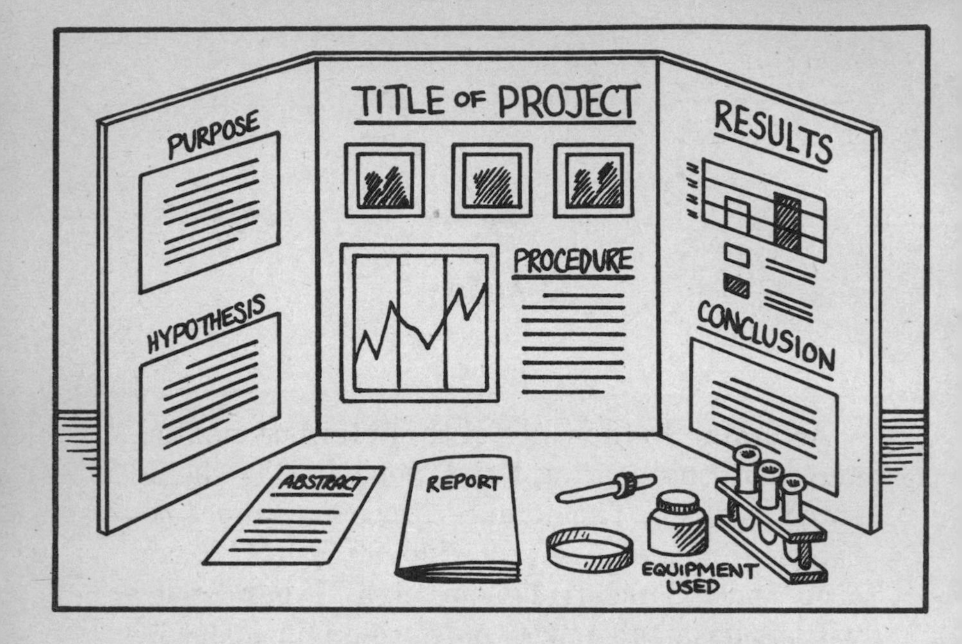

Remember that your physical presentation makes the first impression on a judge. A box with ragged edges and sagging sides may make the judge think you did a sloppy job of research and experimenting as well.

Better ideas for backdrops include hinged wooden frames faced with Masonite or pegboard, sheets of plywood held together with hinges or leather straps, Styrofoam sheets taped together with duct or book tape, Foamkore sheets taped together, and reinforced poster board.

It is important to make your backdrop neat, sturdy, and of sufficient size to display what you've done.

WHAT TO INCLUDE

First, you will want your **Title** to be centered on the middle part of your display board and in large lettering. On the left side you will put your **Purpose.** Below that, you will put your **Hypothesis.** Make these statements short, but complete.

In the center, below the Title, you will have a listing of the steps you followed, your **Procedure.** On the right section, you will show your **Results** (which will be discussed later) and your **Conclusion.** Again, make your statements simple and complete. Be sure your Conclusion refers to the Hypothesis, and that the entire display does indeed reflect the work expected by your title.

On the table in front of your backdrop you will put your **research paper,** any part of your experiment that needs to be displayed (such as plants you grew or a model of a design you tested), and a copy of your abstract. You will not put your name on any of this. Your project will have a number instead.

Check and recheck your display for misspelled words, incorrect titles, and insufficient information. Have someone who is not familiar with the project look over your display to make sure it makes sense.

AESTHETICS

You will want to cover your backdrop with paper, fabric, or poster board. Choose colors that are pleasing as well as eye-catching. If you type your information in small letters on neon raspberry paper, it will certainly be eye-catching, but it will be hard to read. Make sure your writing is easy to read. Make it dark and neat.

You can print the information neatly by hand, generate the writing with a computer and printer, stencil the lettering, or use rub-on or cutout letters. If your hand-lettering is not neat, it would be a good idea to get some help or use another method.

Adding a fabric or colored paper cover beneath your display is a nice touch. Little details in presentation can show just how much care you took with your project. However, even a beautiful presentation cannot save a project if the experiment was done carelessly.

ILLUSTRATIONS, CHARTS, AND DIAGRAMS

In addition to the required parts of the display mentioned above, you may want to include photographs, charts, or diagrams. Remember that what you include in your display should be the most important parts of your work.

If you take photographs of your work, be sure they are clear, and that it is easy to tell what is happening in them. For example, if you measure how far an earthworm travels in one day in a box of soil, you might want to photograph the trail.

Clearly mark the beginning and the end of the trail. It might be a good idea to set a ruler next to the trail to give the observer a reference for length.

You will not be able to bring your earthworm to the science fair. Similarly if you ran a test to see how quickly a mouse could learn to run a maze, you will not bring the mouse. Animals are not allowed, so photographs of them are a helpful substitute.

If you made a machine of some sort, you should photograph or draw a diagram of the inside of your machine. Make your illustrations clear and simple, and label them if necessary.

Don't include photographs of something you can easily bring to the exhibit. If you have two jars of sand with which you experimented, bring the jars themselves. Remember, you want to keep your display simple but also as complete as possible.

GRAPHS AND CHARTS

After you've done your experiment, you'll probably have many numbers or measurements from your work. Some of those numbers will be important to display in a graph or chart. An important thing to remember is that you want anyone who reads it to understand immediately what you did in your experiment.

Label your charts and graphs clearly. Each will need a simple title, telling the observer exactly what the chart is about. You will need an x axis and a y axis, or two perpendicular lines. On the y axis, or vertical line, you will mark and label several segments. For example, if you are comparing percentages, the numbers on your vertical line could include 0, 25, 50, 75, and 100. Always begin with 0 where the two lines meet. Your segments must be equal and of equal value.

Include a legend to name the parts of your graph. Do not include numbers at the top of each bar. Your report should include these numbers. The graph will show the comparison visually.

Don't use fractions. Use decimals. Don't include numbers you don't need. Let's say you compared the average time it took a mouse to run a maze against the average time it took a gerbil. In your report you will chart all of the times. But on your display, you will want to chart just the averages. Your graph will show the difference between the two averages.

9.
Judging

The third part of your presentation is speaking with the judges. Before you talk with them, they will have looked at your display and read your report. Judges will have a list of things to look for in your work. They will rate your project with scores in several areas. A typical rating scale is:

Creative ability: 30 points
Scientific thought: 30 points
Thoroughness: 15 points
Skill: 15 points
Clarity: 10 points

As you can see, originality is very important. Many projects are seen every year at science fairs. Try to use your imagination. Use the projects in this book as a starting point. Think of ways you can do them differently, or think of your own project and follow the basic steps shown in the book. If you do a lot of research, you'll be able to come up with many projects and many variables. The hardest part sometimes is deciding which idea is best. Try to choose the most creative experiment that can be accurately performed and measured.

Scientific thought includes your ability to narrow your topic and your control of variables. If you've done a good job on both your research and your experimentation, and can interpret your results correctly, your score here should be high.

Thoroughness has to do with your intentions and what actually happened. It's okay to discover that your hypothesis is wrong. But it's important that you test what you intend to test and that you do it as completely as possible. If you did your whole project in one day, your score here will be very low. If you've done a lot of research and worked carefully, your score will be higher.

The *skill* section of judging will take into consideration how much help you had. Don't pick a project on nuclear physics if you don't have direct access to the information and equipment you'll need. If someone does the work for you, your score will be low. If you are *helped* by an adult, but you do the work yourself, your score will be higher.

Clarity involves your ability to present your project. When you talk with a judge, try to answer all the questions as clearly and completely as possible. The judge will want to know if you really understand what you did. Can you explain the purpose, hypothesis, procedure, and conclusions? Did you make your presentation clear? Can the judge tell from your display just what you did and what you found out?

It's a good idea to practice talking about your project. Set up your display at home. Ask an adult to ask you questions about your project. Make sure you can explain exactly what you did and what you found out. The more you talk about it, the more relaxed you'll be when you speak to a judge. If you found out something that you haven't mentioned in your research, tell the judge about it. Your excitement about the project will impress the judges.

On the day of the science fair, dress up a bit. Professionals wear their best clothes when they present their findings. You should, too.

Listen carefully to each of the judge's questions. If you don't understand, ask the judge to rephrase the question. Then answer as clearly and completely as possible. If you can't answer a question, tell the judge. Don't try to fake it.

Don't repeat yourself. Avoid slang and don't say, "You know?" Don't chew gum or eat candy.

Speak clearly and smile at the judge. Show the judge just how interested you are in your project. Be enthusiastic. And, don't forget to thank the judge for his or her time.

10.
Projects

WOULD ROCK MUSIC IN THE CLASSROOM INCREASE MY TEST SCORES?

Behavioral Science

PURPOSE: To determine if listening to rock music affects test taking.

MATERIALS: Two tests; class of students; tape of rock music; tape player.

PROCEDURE:

1. Develop your hypothesis.
2. Choose your group of subjects. Check with your teacher. Maybe it would be a good idea to use your science class. The group taking the test should be a fairly homogeneous group. A class of sixth-grade science students at your middle school would be a homogeneous group. The people on a city bus in the middle of the day would probably not be a homogeneous group.
3. Ask your teacher to make out a typical test the class would take.
4. Take half the problems, the odd-numbered ones, and put them together to make one test. Make a second test out of the even-numbered ones.
5. Give the class the first test on one day. Do not play rock music or make any changes from their normal test taking.

6. The next day, give the class the other half of the test. This time, play rock music during the test. Make no other changes from their normal test taking.
   Grade the tests and compile the scores. Compare them, using a chart or graph.

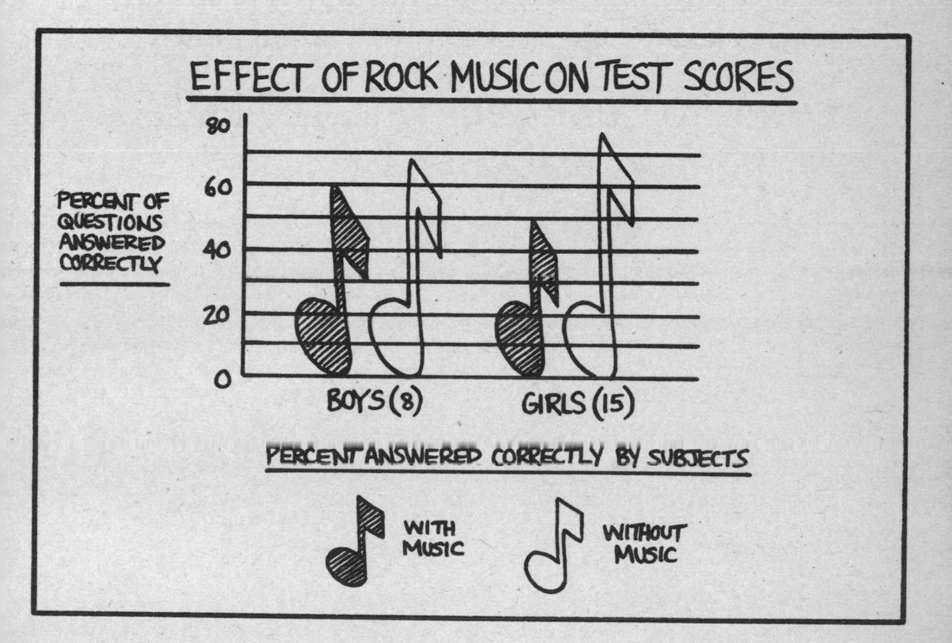

REMEMBER: The only thing you want to be different from one test-taking period to the other is the addition of rock music. However, be aware that the day a test is given sometimes affects the results.

RESOURCES: Check your library for books and articles about concentration, test taking, music.

THINGS TO THINK ABOUT: Could you use two different classes on the same day?

DOES VIEWING TELEVISION AFFECT PULSE RATES?

Behavioral Science

PURPOSE: To determine if pulse rates increase or decrease while viewing TV.

MATERIALS: Television; clock, watch, or stopwatch with a second hand; paper for recording.

PROCEDURE:

1. Develop your hypothesis.
2. Set your target group. Ask your subjects to participate and set a date.
3. For each subject:
   - A. Have the subject sit and rest for at least three minutes.
   - B. Take the subject's pulse for one minute by pressing on the side of the neck or at the bend of their wrist with your index and middle finger. Do not use your thumb. (If you've done your research on taking a pulse, you'll know why.) You have now established the resting pulse rate for this subject. Record it.
   - C. Have the subject sit still and watch television.
   - D. After about fifteen minutes, retake the subject's pulse. Record it.

4. Repeat step 3 for all subjects. Chart your findings.
5. What conclusions can you draw from the data?

REMEMBER: You need to have each subject view the TV separately, eliminating the problem of conversation or other distractions.

RESOURCES: Check your library for books and articles about taking a pulse, pulse rates, stimulation, and the effects of TV.

THINGS TO THINK ABOUT: Could you restructure this experiment using the same show and different subjects, or different types of shows with the same subjects?

DO STUDENTS AT MY SCHOOL PREFER PEPSI?

Behavioral Science

PURPOSE: To determine, by means of a survey, which soft drink is most popular with students at my school.

MATERIALS: Copies of survey; charts; markers.

PROCEDURE:

1. Develop your hypothesis.
2. Determine your target group. Will it be a random sampling of kids you see in the hallway, or just your friends? Will you ask kids of all ages at your school? Will you ask just boys? Just girls? Just sixth-graders? Be sure your results and your title reflect this decision.
3. Make up your survey question. For example, "Which type of soda pop is your favorite?" Don't include a brand name in your question.
4. You might want to go to the supermarket to see all the different kinds of soda pop, including low-sodium, caffeine-free, etc.
5. Make a recording sheet. Have columns for marking "Pepsi," "Coke," "Diet Caffeine-free Coke," etc. Be sure to have a column for "Doesn't know" and "Doesn't drink soft drinks."
6. Practice asking your question to your family

members. Don't count these answers since they're only for practice.

7. Ask as many schoolmates as possible. Note every answer under the appropriate column.

REMEMBER: Ask each person just one time. Tell the person you are taking a survey for a science project. Try not to influence the answers. Make your question very specific. Many people will just answer "Coke" or "Pepsi," so ask specifically which type (Diet, Caffeine-free, etc.) they drink.

RESOURCES: Check your library for books and articles about health and the effects of caffeine on the body. Look at the cans themselves. Some companies have toll-free numbers you can call for information about the products.

THINGS TO THINK ABOUT: Could you ask your subjects which are their top three favorite soft drinks, and have them rank them? Give their favorite three points, the second two points, and the third one point. This would give you more data to work with and analyze.

DOES PLANT FOOD REALLY HELP MARIGOLDS?

Botany

PURPOSE: To determine if the root system of a marigold plant is greater, by weight, when the plant has been watered with just tap water or with plant food solution.

MATERIALS: Six peat pots; soil; plant food solution; water; marigold seeds; charts.

PROCEDURE:

1. Develop your hypothesis.
2. Plant three marigold seeds in each of the six peat pots. Label three TAP WATER. Label the other three PLANT FOOD SOLUTION.
3. Water three pots with tap water and the other three with a plant food solution, according to the labels. Put the pots in a sunny location. Add water as needed to keep the soil moist.
4. Thin down the seedlings to one plant per pot when they are two inches tall. Continue to care for the plants until you near the science fair date.
5. Take the plants labeled PLANT FOOD SOLUTION out of the pots. Carefully rinse the soil from the roots.
6. Cut off the tops.

7. Weigh and record the weights of the roots.
8. Repeat steps 5–7 with plants labeled TAP WATER.
9. Compare your results.

REMEMBER: Do everything exactly the same for each group except watering with tap water or plant food solution. Be sure to give yourself enough time for the plants to grow several inches tall. The longer you can run this experiment, the more impressive your results will be.

RESOURCES: Check your library for books and articles about root systems, fertilizer, plant growth, marigolds.

THINGS TO THINK ABOUT: Keep the tops of your marigolds. Could you design a project measuring the differences in them?

DOES MAGNETISM AFFECT PLANT GROWTH?

Botany

PURPOSE: To determine the effects of magnetism on zucchini plants.

MATERIALS: Four zucchini bedding plants; four pots; potting soil; four small bar magnets.

PROCEDURE:

1. Develop your hypothesis.
2. Place all four zucchini bedding plants in pots with potting soil.
3. Write the numbers 1, 2, 3, and 4 on four pieces of paper. Put them in a bowl and pull out a number for each plant. This guarantees a random choice for both sides of the experiment. Label the pots accordingly.
4. Add two bar magnets to the soil of plants 1 and 2, according to the diagram on the page that follows.
5. You must decide BEFORE you begin what you are going to compare. The cotyledons or leaf formation? The weight of the plants? The height of the plants? The color of leaves? The appearance of the first true leaves?
6. Care for all the plants in the same manner. Make notes in your project notebook about the

type of water, fertilizer, and light you give the plants.

7. Before the science fair, compare the plants. Notice any differences between the two groups — in size, color, etc.
8. Make notes of the growth in your project notebook. Chart the differences you note.

REMEMBER: You will want to make sure the plants receive the same amount of water, sunshine, etc., as much as possible. These are variables you will want to control.

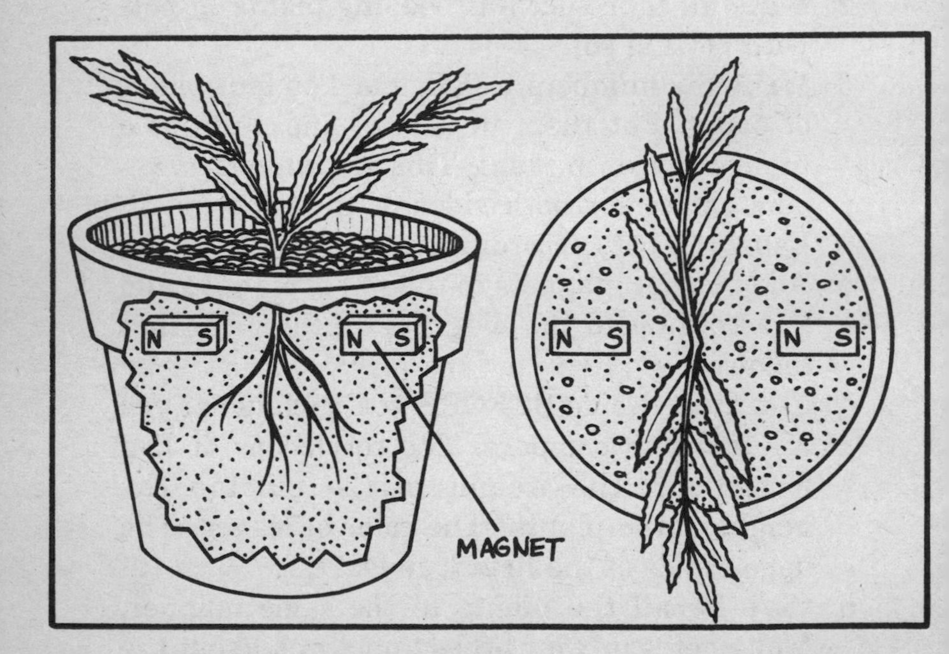

RESOURCES: Check your library for books and articles about magnetism, the effects of magnetism on plants, plant growth, zucchini.

THINGS TO THINK ABOUT: Could you design an experiment that would add a force different from magnetism? Talk it over with your teacher or a botanist.

HOW DOES HEAT FROM BELOW AFFECT ROOT DEVELOPMENT IN HOUSEPLANTS?

Botany

PURPOSE: To determine what effects, if any, heat from below has on root development.

MATERIALS: Four small houseplants; thermostatically controlled electric warming tray (UL listed *only*).

PROCEDURE:

1. Develop your hypothesis.
2. Buy four small, easy-to-grow houseplants, such as pothos, at a nursery. Make sure they all are healthy and the same size and type. Photograph the plants.
3. Mark the pots of two of the four plants and set them in a cool room on a warming tray in a location that offers the plants light.
4. Set the other two plants in the same location, but *not* on the warming tray.
5. Care for the plants according to the nursery directions until close to the date of the science fair.
6. Carefully wash the soil from the roots of the plants.

7. Compare the roots of the plants on the tray with those of the plants not on the tray. What are the differences?
8. Photograph your plants with their roots exposed.
9. Cut the tops off the plants and weigh the roots, in grams.

REMEMBER: Buy the same type of plants to control the variables in your experiment. Be sure to care for each plant in the same way, watering when it needs it. Because of the warming tray, the amount of water each plant needs may vary.

Start this experiment at least a month before the science fair, to give yourself time to have measurable effects.

RESOURCES: Check your library for books with information about plants, root development, the effects of temperature on growth, and what plants need to grow.

THINGS TO THINK ABOUT: If you do not have a warming tray, you could make a solar greenhouse for your plants by building a wooden box with a clear plastic cover. By keeping the cover over the plants, you will be warming the soil. Remember that you need to compare the plants with warm soil to the plants without warm soil. How would you set up a control for plants whose soil is not warmed?

How can this help you in gardening?

Can you design a project that does not kill the plants when you are finished?

HOW MUCH CAFFEINE DO I GET WHEN I DRINK A SODA POP?

Medicine and Health

PURPOSE: To find the amount of caffeine in various brands and types of soda pop.

NOTE: This is a research project.

MATERIALS: Many references; chart paper; markers.

PROCEDURE:

1. Make a list of all the kinds of soda pop you want to test. Be sure to include many different brands and types.
2. Check the cans or bottles to see if they have information available.
3. Write to the soda pop companies. Explain what you are doing and request more information.
4. Chart your findings.
5. What conclusions can you draw from the data you've gathered?

REMEMBER: The more you put into a research project, the better it will be.

Make sure you present your information in standard units. For example, tell how much caffeine is in a twelve-ounce serving for each soda pop, not how much is in one ounce of one kind, and in sixteen ounces of another. Remember to check for serving size and number of servings on the container.

RESOURCES: Check your library for books and articles about caffeine, soda pop, soft drinks.

THINGS TO THINK ABOUT: Are there other names for caffeine that might be listed in the ingredients? Are there other chemicals added to the soda pop that might influence the person who drinks it? Could studying these chemicals make a good project, as well?

IS MY DOMINANT HAND STRONGER?

Medicine and Health

PURPOSE: To determine the difference in fatigue level between a subject's right and left hands.

MATERIALS: Small ball for squeezing; a clock, watch, or stopwatch with a second hand; paper for taking notes.

PROCEDURE:

1. Develop your hypothesis.
2. Have each subject:
   A. Squeeze the ball as tightly as possible in the right hand and hold one second. Repeat until the right hand feels fatigued. Note the time it took for the hand to become fatigued. Write it down.
   B. Repeat step A for the left hand. Note the time.
   C. Note whether the subject is right-handed or left-handed.
3. Compile the times for several right-handed and left-handed people. Compare the numbers. Were right-handed people able to squeeze the ball more times with their right or left hands? Left-handed people?

REMEMBER: You will want your subjects to

quit squeezing when they feel the same amount of fatigue in each hand. Be sure your subject group includes some right-handed and some left-handed people.

RESOURCES: Check your library for books and articles about muscles, muscle fatigue, right-handedness and left-handedness.

THINGS TO THINK ABOUT: Could you design an experiment to strengthen your nondominant hand?

DOES CAFFEINE IMPROVE ATHLETES' REFLEXES?

Medicine and Health

PURPOSE: To determine the effect of caffeine on a middle-school athlete's reflexes.

MATERIALS: Strips of paper, 5 cm × 30 cm, marked every cm; caffeinated drinks; clock.

PROCEDURE:

1. Develop your hypothesis.
2. Obtain volunteers from a group of middle-school athletes to be your subjects. Ask them not to eat or drink anything containing caffeine (such as some kinds of soda pop, coffee, tea, chocolate, and some pain relievers), for several hours before your test.
3. Test the reflexes of your subject before exposure to caffeine.

TEST: Hold a paper strip just above your subject's hands. Have the subject hold out his or her hand with thumb opposing fingers. (See the diagram on the following page.) Tell the subject to catch the paper strip between thumb and index finger as you drop it. Measure where the subject grips the paper. Record this number.

4. Immediately repeat the test five times for each subject. Record all test results.
5. Give half your subjects a 12-ounce drink con-

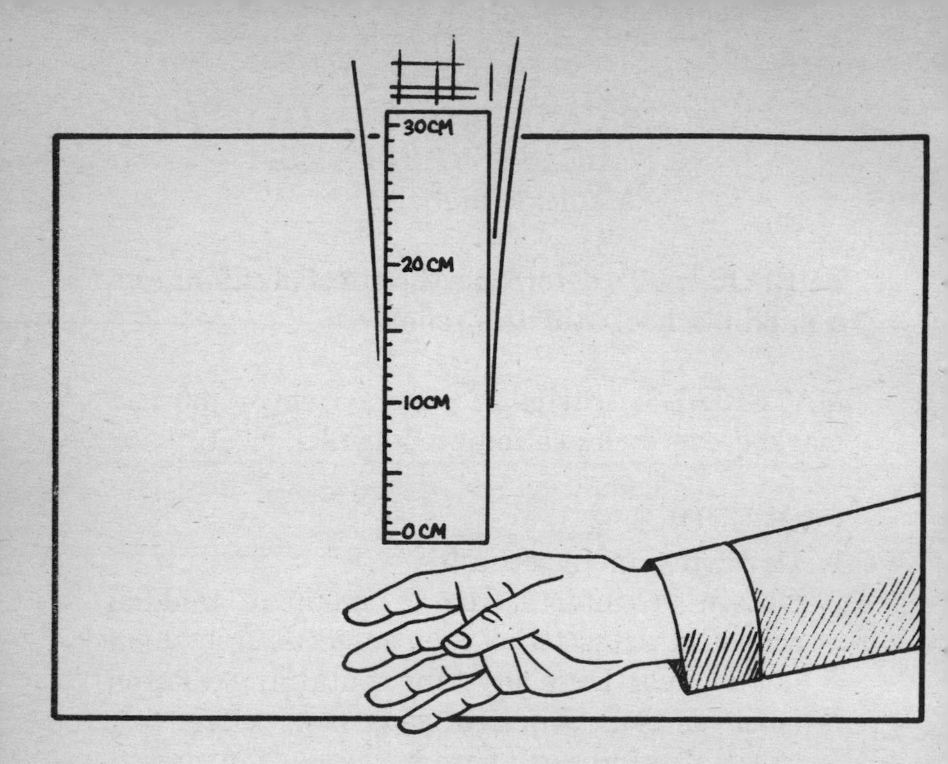

taining caffeine. Give diet, caffeine-free soda to the other half. Repeat the test, five times, when you think the caffeine has taken effect. You will know about how long it takes after you have done your research on the effects of caffeine.

6. Have the subjects drink a second 12-ounce drink, the same as their first. Repeat this test five times in thirty minutes.
7. Compare the results.

REMEMBER: Record all results accurately. The tests for each subject must be given during a single time period. Let your subject know how long this test will take.

RESOURCES: Check your library for books and articles about reflexes, caffeine, stimulants.

THINGS TO THINK ABOUT: How will you know if the effects you obtain are caused by caffeine? Could they be caused by sugar? Could you control the sugar variable by using diet soda?

HOW DOES SMOKING AFFECT PULSE RATES AFTER EXERCISE IN ADULTS?

Medicine and Health

PURPOSE: To determine if smoking affects the recovery rate of pulse after exercise in middle-aged adult males.

MATERIALS: Paper for recording; a clock, watch, or stopwatch with a second hand.

PROCEDURE:

1. Develop your hypothesis.
2. Obtain volunteers who are middle-aged adult males. Half your subjects should be smokers and half nonsmokers.
3. Take the pulse rate of each subject after at least three minutes of sitting and resting. Record it.
4. Have each subject jog in place for two minutes, then stop.
5. Now take the subject's pulse for one minute. Record it. Start the stopwatch to mark the time.
6. Have the subject sit and rest for two minutes.
7. Repeat steps 5 and 6 until subject's resting pulse rate is regained. How long did it take? Write down the time.
8. Compare the results of the two groups.

REMEMBER: Keep accurate records.

RESOURCES: Check your library for books and articles about smoking and heart rates, pulse rates. Also check with the American Heart and Lung Associations for information.

THINGS TO THINK ABOUT: In spite of the surgeon general's warnings, people continue to smoke. How could your project help someone make the choice to stop smoking or not to start smoking?

WHAT VITAMINS ARE IN SCHOOL LUNCHES?

Medicine and Health

PURPOSE: To determine the vitamins contained in the lunches served in my school during one week.

NOTE: This is a research project.

MATERIALS: Lunch menu; nutritional information; chart paper.

PROCEDURE:

1. Ask the manager of your school cafeteria to help you. List the items served for lunch for each day in the week for one week. Make sure you include serving sizes.
2. You may need to have the manager help you include ingredients for certain things, such as the amount of meat, cheese, lettuce, etc., for a taco.
3. List all the ingredients for each day.
4. Using your resources, find the vitamin content of each item.
5. Calculate the vitamin content for each day. Chart the vitamins for each day. Then average the vitamin content for a week. Compare it to the recommended daily allowance for each vitamin.

REMEMBER: You will need to take one week's menus. Don't take menus from two different weeks unless you change your Purpose. Maybe you'd like to analyze the vitamins offered on Mondays versus the vitamins offered on Fridays for a month.

RESOURCES: Check your library for books and articles about vitamins, nutrition, recommended daily allowances for vitamins.

THINGS TO THINK ABOUT: It you are in elementary school and eat cereal for breakfast, you might want to analyze the vitamin content of your breakfast instead of lunch, as vitamin information is readily available on cereal boxes.

ARE A DOCTOR'S HANDS REALLY CLEAN?

Microbiology

PURPOSE: To determine the effectiveness of using various soaps to destroy surface bacteria on skin.

MATERIALS: Three different types of soap, including one used by a surgeon before an operation, one used by a doctor in his or her office, and one used in the home; five agar plates.

PROCEDURE:

1. Develop your hypothesis.
2. Turn one agar plate upside down without opening it to make sure your agar is not contaminated. If nothing grows, your agar is pure and you'll be able to assume anything that grows in your other dishes is from the subject's hands. Label it.
3. Culture the subject's hands by touching the agar. Cover the agar plate and turn it upside down to keep moisture from affecting the bacteria. Label it. This is your control.
4. Have subject wash hands with one of the soaps for one minute. Repeat step 3.
5. At the same time the next day, have the subject repeat step 4. Use a different soap.

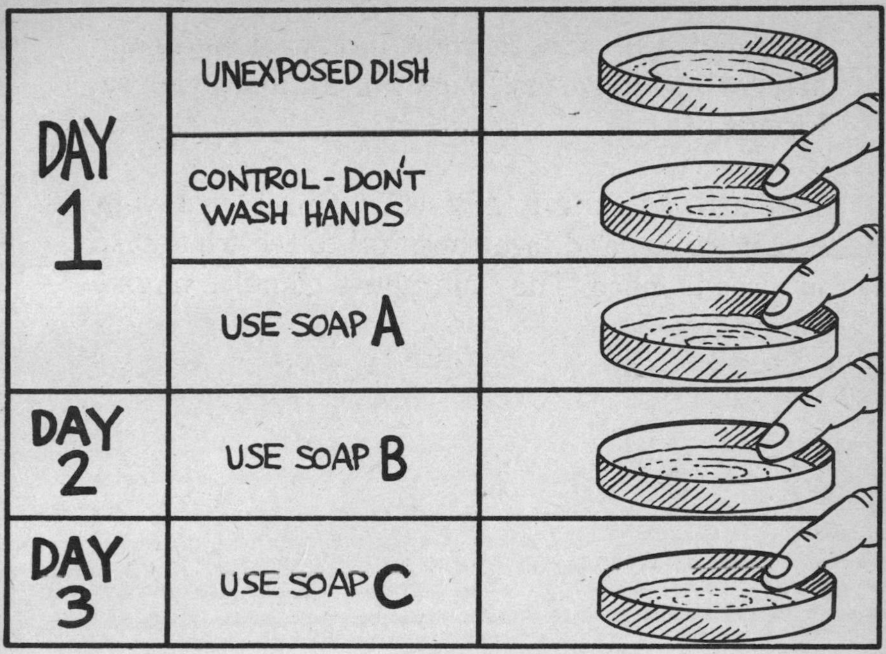

DAY 1	UNEXPOSED DISH
	CONTROL - DON'T WASH HANDS
	USE SOAP A
DAY 2	USE SOAP B
DAY 3	USE SOAP C

6. Again, the third day, have the subject repeat step 4, using the third soap.
7. Compare what grows in each plate. Which soap was most effective? Least effective?

REMEMBER: Culture the subject's hands at the same time each day. For example, after school and *before* subject washes his or her hands. Label each agar plate carefully. Be careful when handling the plates. An unwanted thumbprint may skew your results. Ask your science teacher about the proper way to dispose of the agar plates.

RESOURCES: Check your library for books and articles about germs, bacteria, antibacterial solutions.

THINGS TO THINK ABOUT: Check with your doctor, clinic, and local hospital to see what soap is recommended. They may have samples to give you to test.

WHY SHOULD I WASH MY HANDS?

Microbiology

PURPOSE: To show the difference in growth of germs in a culture taken from a subject's hands before and after washing.

MATERIALS: Seven agar plates; soap.

PROCEDURE:

1. Develop your hypothesis.
2. Turn one agar plate over without uncovering it. This is to make sure your agar is not contaminated, which would ruin your findings. Label it.
3. Remove bacteria from the subject's hands by having the subject touch a fresh agar plate. Label the agar plate and turn it upside down to keep moisture from affecting the bacteria.
4. Have subject wash hands with soap and water.
5. Repeat step 3 with a fresh agar plate.
6. Repeat these three steps two more times, on different days. You will have three plates cultured by unwashed hands, and three cultured by washed hands, as well as one plate that you didn't touch.
7. Compare the bacteria cultured in each plate. When was more bacteria grown, before or after washing hands?

REMEMBER: Label each agar plate carefully. Be careful when handling the plates. An unwanted thumbprint may skew your results. Ask your science teacher about the proper way to dispose of the agar plates.

RESOURCES: Check your library for books and articles about the growth of bacteria, germs, health practices, hands.

THINGS TO THINK ABOUT: Are your hands cleaner at different times of the day? Could you design an experiment to see?

IS THERE REALLY A DIFFERENCE AMONG BRANDS OF DETERGENT?

Chemistry

PURPOSE: To determine which detergent removes the most dirt from dirty socks when they are washed in cold water.

MATERIALS: Four pairs of white socks, three different brands or kinds of detergent (for example, Bold liquid, Tide powder, ERA); washing machine; cloth labels for socks; indelible marker.

PROCEDURE:

1. Develop your hypothesis.
2. Put one sock in a jar with muddy water and grass. Shake for one minute. Do the same for all eight socks.
3. Label the socks: one pair for each type of detergent, plus one pair for no detergent as a control. Wash each pair of socks in a different detergent and cold water in the washing machine. Use the same setting and length of wash.
4. Dry socks either outside or in the dryer. Dry all socks in the same way.
5. Compare the socks. Put each sock against a white piece of paper. Which sock is most white? Which is next? Which is the least white?

REMEMBER: The detergent should be your only variable. Be sure to dirty, wash, and dry the socks in the same way.

RESOURCES: Check your library for books and articles about detergent, stains, biodegradability.

THINGS TO THINK ABOUT: A variation on this experiment would be to cut fibers from the same place on each sock after laundering and drying. View the fibers under a microscope. Then make your comparisons.

Which detergent is the best for cleaning socks and having the least negative impact on our environment?

CAN A NAIL BE MADE RUSTPROOF?

Chemistry

PURPOSE: To determine if coating nails will prevent rust from forming on them.

MATERIALS: Nails; various materials for coating such as petroleum jelly, motor oil, paint (both latex and oil base); a tray smaller than a cake pan; a cake pan; plastic wrap; a cloth (optional).

PROCEDURE:

1. Develop your hypothesis.
2. Make sure you identify the type of nails you are using. Use all the same kind. Try using iron nails that are not coated. Ask for help in the hardware store where you purchase your nails.
3. Line your tray with plastic or cloth. Coat each nail in one type of material only. Let the coating dry. Label the spot on the tray.
4. Have one nail that is not coated for a control.
5. Place the tray in a cake pan. Pour water around the tray in the cake pan. Cover the top with plastic wrap. (See diagram on the next page.)
6. Let the nails sit in the humid place for about a month. You may have to add more water to the cake pan. Don't let it dry up. Then compare the rust that has formed on the nails. You should check for both visible and hidden (under the coating) rust.

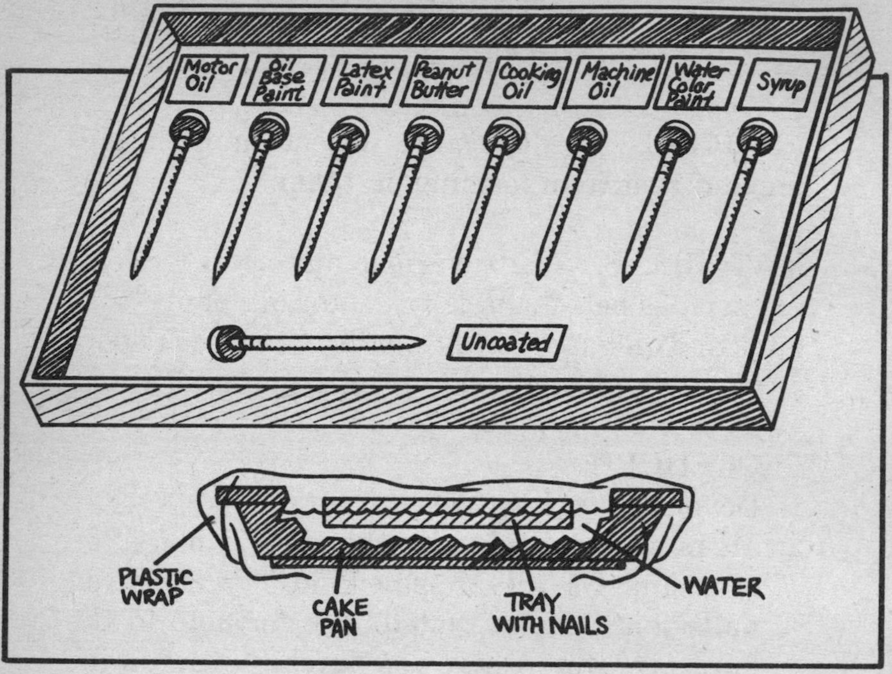

REMEMBER: You will need to compare the rust formed on each coated nail with the rust on the control (uncoated) nail.

RESOURCES: Check your library for books and articles about rust, iron, metals.

THINGS TO THINK ABOUT: A variation of this project might be to determine the type of nail least likely to rust in a humid environment by trying to rust many different types of nails.

DO FROZEN LIQUIDS MELT IN THE SAME AMOUNT OF TIME?

Chemistry

PURPOSE: To determine whether various kinds of frozen liquids melt in the same amount of time.

MATERIALS: Twelve liquids (water, carbonated beverage, milk, shampoo, orange juice, etc.); measuring cup; twelve plastic cups; spoons or popsicle sticks; cookie sheet or muffin tin to hold cups; freezer; clock; record sheet.

PROCEDURE:

1. Develop your hypothesis.
2. Pour the same amount of each liquid into the cups.
3. Put the cups on a cookie sheet or in the muffin tin to hold them. Add a spoon or popsicle stick to each cup of liquid, and put the cups in the freezer. Let the liquids freeze solid, at least overnight.
4. Remove the cups from the freezer. Record the time on your sheet. Try each spoon or stick in turn. Try again every few minutes. When the spoon can be pulled from the liquid, record the time on your sheet. Then mark the time it takes for each liquid to thaw.

REMEMBER: Don't worry if your experiment

causes you to reject your hypothesis. Scientists often learn a great deal from experiments that prove their guesses wrong. The experiment and the proof are the important part.

RESOURCES: Look for books in the library about liquids, freezing, ice formation, and temperature.

THINGS TO THINK ABOUT: Try measuring the difference in the time it takes for various flavors and types of ice cream, ice milk, and sherbet to melt. Why does ice cream taste so good on a hot day?

CAN I MAKE A BIGGER BUBBLE?
Chemistry

PURPOSE: To determine the effect of glycerin on the size of soap bubbles.

MATERIALS: Glycerin (can be found in drug stores); two identical bubble makers (a good bubble maker is a large, 10-inch, plastic macramé hoop found in craft stores); two large pans (larger than your bubble maker); bubble solution made from 125 ml dish detergent (Dawn or Joy work well) and 4 liters water.

PROCEDURE:
1. Develop your hypothesis.
2. Make bubble solution. Divide evenly between two pans.
3. Add 45 ml gylcerin to one of the pans. Label both pans.
4. Dip the bubble maker in one pan and pull it through the air to make bubbles.
5. Rinse the bubble maker with tap water.
6. Repeat step 4 using the second pan of solution.
7. Compare the bubbles.

REMEMBER: You will want to make a lot of bubbles to compare the size of them. Photographing them would be a good idea. The photos would document your results. You will need a helper for this. Go outside to make your bubbles or else you will have a slippery mess.

RESOURCES: Check your library for books and articles about bubbles, glycerin, water tension.

THINGS TO THINK ABOUT: Experiment with the amount of glycerin you add to find just the right amount to make the largest, longest-lasting bubbles. Try creating your own bubble makers to make huge bubbles, many tiny bubbles, or bubbles within a bubble, etc.

WHAT DOES SOAP DO TO WATER TENSION?

Chemistry

PURPOSE: To determine if the addition of soap increases or decreases the weight necessary to break water surface tension.

MATERIALS: A balance (or make one out of a milk carton, a wooden dowel, two flat pans, and string, thread, or fishing line — see diagram on following page); weights; pan of water; liquid dishwashing detergent; flat pan.

PROCEDURE:

1. Develop your hypothesis.
2. Set up your balance. The left side of the balance will hold weights. The right side should be a flat pan smaller than your pan of water.
3. With no weights added on to the left side, the water level in your pan should be just above the level of the right side.
4. Add weights to the left side until the right side is lifted free of the water. Record how much weight it took to lift the pan. Repeat this five times.
5. Add one teaspoon of liquid dishwashing detergent to the water. Stir gently.
6. Repeat step 4.

7. Compare the results. Ask your mathematics teacher to help you with the statistics.

REMEMBER: You will want to repeat this several times and average the weights. Make sure when you repeat it to rinse the pan well and use fresh water when you begin.

RESOURCES: Check your library for books and articles about water tension.

THINGS TO THINK ABOUT: Ask your science teacher if you can borrow a set of weights that measures in grams. Can you think of other things you can use as weights? Pennies, for instance?

DOES SALT AFFECT THE FREEZING PROCESS OF WATER?

Chemistry/Physics

PURPOSE: To determine if the addition of salt increases or decreases the time needed to freeze water.

MATERIALS: Salt; distilled water; identical cups; measuring spoons; spoons; freezer.

PROCEDURE:

1. Develop your hypothesis.
2. Fill a cup with 200 ml of distilled water. Label it, DISTILLED WATER.
3. Fill an identical cup with 200 ml of distilled water. Add 5 gm salt. Stir. Label this cup, WATER AND SALT.
4. Using two more cups, repeat step 3, but put 2 gm salt in one cup, and 10 gm salt in the other cup.
5. Put a spoon in each cup and put the cups in the freezer compartment of your refrigerator.
6. Mark down the time.
7. Check on the cups every fifteen minutes until they are frozen and the spoons cannot be pulled out from the frozen liquid.
8. Mark the time it took for each cup to freeze solid.
9. Chart your results.

REMEMBER: Keep careful track of the time it took each cup to freeze solid. Repeat the experiment several times and average the times.

RESOURCES: Check your library for books and articles about salt, freezing water, temperature.

THINGS TO THINK ABOUT: Do other substances affect the freezing process of water? Why do some cities pour salt on icy streets? What else do they use that doesn't hurt plant life?

DO BRAND AND TYPE OF BATTERY REALLY MAKE A DIFFERENCE?

Chemistry/Physics

PURPOSE: To determine which of three battery types lasts longest in a tape player.

MATERIALS: Three types of batteries, such as alkaline, carbon, rechargeable; tape player; tapes; clock.

PROCEDURE:

1. Develop your hypothesis.
2. Make sure each battery is new and/or freshly charged.
3. Keep accurate records. Put tapes into your tape player and turn it on. Let it play until the batteries die. Note the time it took to use all the batteries' charge.
4. Repeat step 3 for each type of battery.
5. Compare your results. Make charts.

REMEMBER: Do this experiment at a time when you will be around to know when the tape player stops. Keep notes in your notebook about the quality of sound near the end of the battery. It will not affect your answer to the question, but will be interesting.

RESOURCES: Check your library for books and articles about batteries, electricity, tape players.

THINGS TO THINK ABOUT: Which battery is the most environmentally sound?

DOES COLOR AFFECT REFLECTION OF HEAT?

Physics

PURPOSE: To determine whether colored paper behind an ice cube affects the speed of its melting.

MATERIALS: Measuring cup; small plastic cups; colored paper; chart paper; clock; freezer.

PROCEDURE:

1. Develop your hypothesis.
2. Measure equal amounts of water into small cups and freeze to make identical-sized ice cubes.
3. Place colored paper in the sunshine. Choose a variety of colors and intensities.
4. Place an ice cube on each sheet of colored paper.
5. Mark the starting time.
6. Check each ice cube every thirty seconds until they have all melted. Mark on your chart the time it takes each one to melt completely.
7. Compare your results.

REMEMBER: Don't use so many ice cubes and colors that you can't keep track of them all. Make sure not to block the sun on any of the papers. Take photographs to document your work and add to your display.

RESOURCES: Check your library for books and articles about heat and cold, reflecting and absorbing heat, melting.

THINGS TO THINK ABOUT: Could you design a project using something other than ice melting to measure the heat reflection?

DOES HOT WATER FREEZE FASTER THAN COLD WATER?

Physics

PURPOSE: To determine if hot water freezes at a different rate from cold water.

MATERIALS: Ice-cube trays; sticks; hot water; cold water; clock.

PROCEDURE:
1. Develop your hypothesis.
2. Measure and pour hot tap water into one ice-cube tray.
3. Measure and pour an equal amount of cold tap water into another, identical, ice-cube tray. Add a stick to one compartment of each tray.
4. Put them in the freezer and note the time. Do not stack them on top of each other.
5. Keep checking on them every 15 minutes at first and then every few minutes as they begin to freeze, until the ice is frozen hard and the sticks cannot be removed. Note the time each freezes completely.
6. Make a chart.
7. Compare your results.

REMEMBER: Put the ice-cube trays in similar areas of the freezer. It would be a good idea to repeat the experiment more than one time.

RESOURCES: Check your library for books and articles about ice, heat, cooling, freezing, water currents.

THINGS TO THINK ABOUT: Why would you want to know the answer to this? How can you use the results to save energy in your home?

CAN I MAKE A CURVE USING STRAIGHT LINES?

Mathematics

PURPOSE: To show how a curve can be formed using only straight lines.

MATERIALS: Sturdy paper; markers; colored thread; needle; ruler.

PROCEDURE:

1. Mark a horizontal x axis and a vertical y axis on a sheet of paper. Make each 10 cm long.
2. Mark each axis every 1 cm. Number them from 0 to 10.
3. Starting at the top of y, at 10, draw a line from $y = 10$ to $x = 1$. (See diagram on following page.)
4. Then draw a line from $y = 9$ to $x = 2$.
5. Continue in this pattern until you have connected all the points on the y axis with those on the x axis.
6. Can you name the shape you made?
7. Try using lines drawn at different angles.
8. Practice drawing the lines, then try sewing them with colored thread.

REMEMBER: You will need to understand what you are doing in a mathematical sense. Your re-

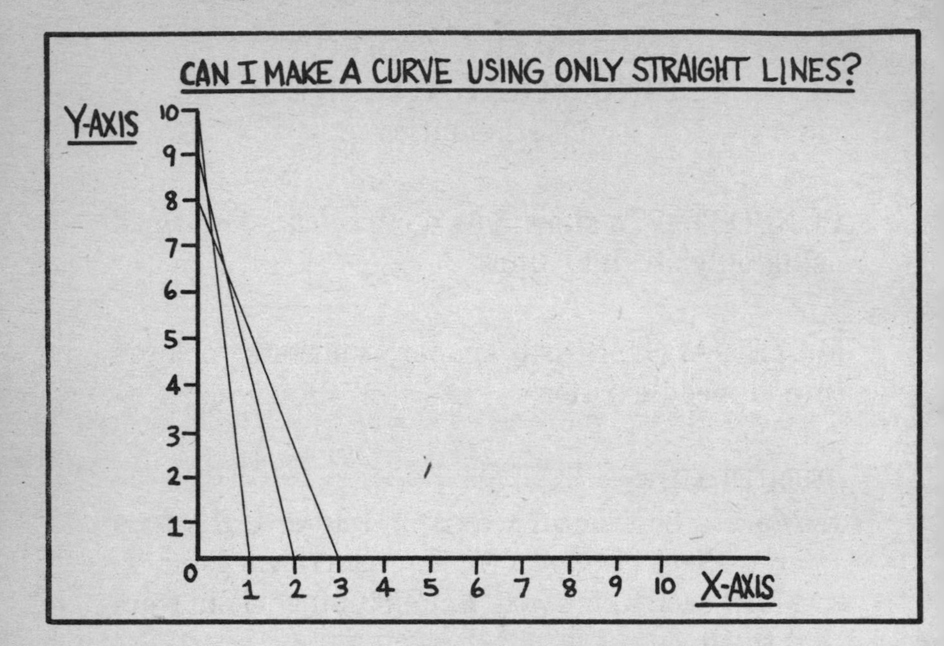

search into parabolas and curved stitching will be important.

RESOURCES: Check your library for books and articles about parabolas, curves, and curved stitching.

THINGS TO THINK ABOUT: Can you develop a computer program to draw the lines?

WHAT IS THE PROBABILITY OF TWO PEOPLE HAVING THE SAME BIRTHDAY?

Mathematics

PURPOSE: To determine the mathematical probability of two people in a group of twenty-five having the same birthday.

MATERIALS: Chart paper; markers.

PROCEDURE:

1. Work out the probability of two people in a group of two having the same birthday.
2. Work out the probability of two people in a group of three having the same birthday.
3. Continue to work out the probability of two people in a group having the same birthday, adding one person to the group each time until you have a group of twenty-five.
4. Make a chart to show what you did and how probability works.

REMEMBER: If you choose to do a project in mathematics, you should work with a mathematician or math teacher, or have a very good understanding of math. The important thing in a mathematics project is your understanding and ability to explain what you did.

RESOURCES: Check your library for books and articles about probability, statistics.

THINGS TO THINK ABOUT: Why would you want to know about probability? Could it help you in your everyday life?

CAN I WRITE A COMPUTER PROGRAM TO DRAW PICTURES?

Mathematics

PURPOSE: To write a computer program that will direct the computer to draw a picture.

MATERIALS: Computer; computer programming guide; chart paper.

PROCEDURE:

1. Think of a simple picture you would like to draw. Use commands you've learned from the programming guide to draw your picture.
2. Use loops to make your drawing faster.
3. Use commands to make your character move.
4. You can add music to your program.
5. Chart your work.

REMEMBER: Your understanding of what you are doing is very important in this project. If you do not have access to a computer, this is not a good project for you.

RESOURCES: Check your library for books and

articles about computers, graphics, computer programming.

THINGS TO THINK ABOUT: Many computer games have sophisticated graphics, or pictures. Could you write a program to play a simple game?

WHICH IS A BETTER INSULATOR, WOOD OR STYROFOAM?

Engineering

PURPOSE: To determine if wood or Styrofoam protects ice from melting longer.

MATERIALS: 5-cm-square sheets of Styrofoam and wood of the same thickness; glue; ice; clock; record sheet.

PROCEDURE:

1. Develop your hypothesis.
2. Make one wooden and one Styrofoam box.

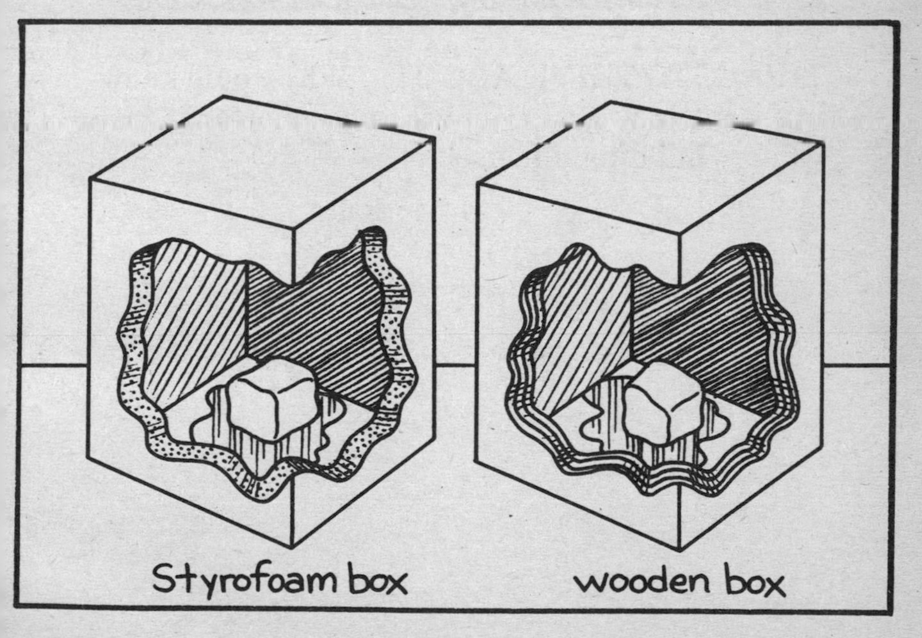

3. Add identical sized chunks of ice and close the boxes. Keep them in the same place.
4. Check on the ice every twenty minutes until it starts to melt. Then check it every five minutes. Record your observations.
5. Note the time that it takes for the ice to melt completely in each box.
6. Compare the results.

REMEMBER: Check quickly and be sure you expose the inside of each box the same amount of time as you check.

RESOURCES: Check your library for books and articles about insulation, heat, heat loss, cold.

THINGS TO THINK ABOUT: Why would knowing what is a good insulator be important when you are building a house?

WHICH PAN COOKS FASTEST?
Engineering

PURPOSE: To determine which type of pan allows water to come to a boil most quickly.

MATERIALS: Several types of pans, all with the same circumference; measuring cup; water; marbles; stove; clock; record sheet; paper and markers for chart.

PROCEDURE:
1. Develop your hypothesis.
2. Make a record sheet for the pans you have gathered.
3. Measure 250 ml of cool tap water and pour it into one pan. Add enough marbles (at room temperature) to cover the bottom of your pan. The marbles will help to keep the boiling even.
4. Put the pan on the stove and turn on the heat to medium. If your stove is a gas stove, make sure you turn the heat to the same mark each time.
5. Time how long it takes for the water to come to a full boil. Record the time on your record sheet.
6. Repeat steps 2, 3, and 4 for each pan.
7. Make a chart showing your results.

REMEMBER: You must do this experiment with an adult. Boiling water can burn you very quickly. Have the adult dump out the boiling water. NEVER touch the hot pan, hot water, or hot marbles. Never pour cold water into a hot pan. Steam can burn, too.

Make sure your pans have the same size diameter so the heating surface will be the same. If you're using very large pans, you might want to try boiling two cups of water.

If you're using an electric stove, don't start boiling the second pan until the heating element has cooled, or use a different burner that is the SAME SIZE. Make sure you begin with cool tap water each time.

Borrow pans from your friends and neighbors to try as many different types of pans as possible, such as cast iron, aluminum, stainless steel, copper-clad steel, glass, etc.

You will want to repeat this experiment a number of times to have more accurate results.

RESOURCES: Check your library for books containing information about heat transfer, cooking, and temperatures for boiling.

THINGS TO THINK ABOUT: Which pan uses the least amount of energy to boil the water? What could this mean for a nation interested in conserving energy?

WHAT SHOES SHOULD I WEAR WHEN I SKATEBOARD?

Engineering

PURPOSE: To determine the type of shoe that is best for controlling a skateboard through specific maneuvers.

MATERIALS: Several types of shoes, such as deck shoes, sneakers, flip-flops, leather-soled shoes; skateboard; butcher paper; sidewalk.

PROCEDURE:

1. Develop your hypothesis.
2. Put on the first pair of shoes and take your skateboard outside.
3. On the butcher paper, draw a right angle. Tape the paper to the sidewalk. Try to follow the angle on your skateboard. After each attempt, trace the wheels' track with a different color. The weight of your body and the skateboard should make a track you can see on the paper. Try each angle with the same shoes several times. Measure the distance from the actual curve for each attempt. Average these numbers. (See diagram on following page.)
4. With new pieces of paper but the same angle, repeat step 3 with each pair of shoes.
5. After each try, write down in your project notebook what happened and how it felt. Pay at-

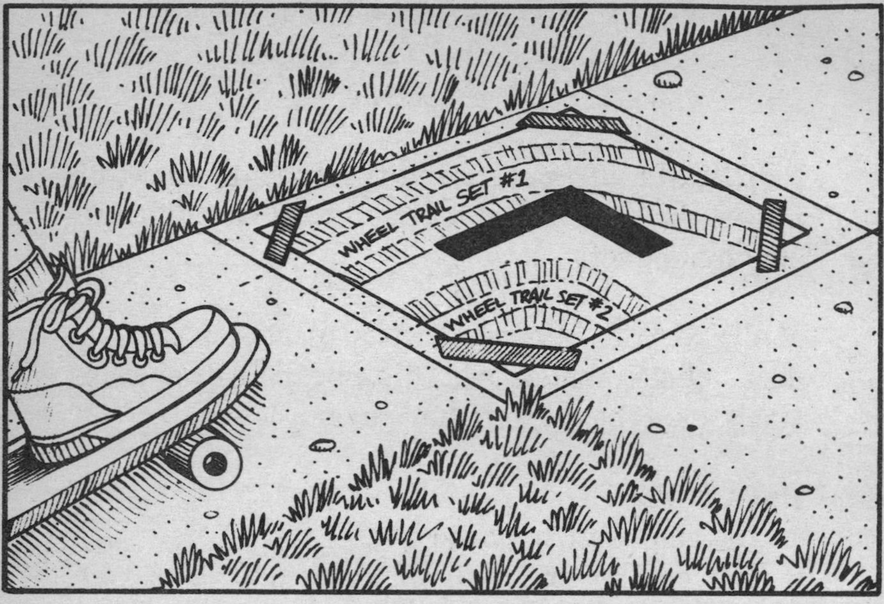

tention to what your feet do in each pair of shoes. Do they slip? Do they cling to the board?

6. Have your friends try. Tell them what to do. Watch their feet. Do the shoes slip? Do their feet slip in the shoes? Measure their accuracy with the right angle.

REMEMBER: Be sure to try every maneuver with each pair of shoes. Keep an open mind. Just because you like a particular shoe, don't assume it will be the best. Action photographs of the shoes on the boards would be good.

RESOURCES: Check your library for books and articles about shoes, friction, rubber and leather soles.

THINGS TO THINK ABOUT: Could you use the information you've gained to pick the safest design of shoes for skateboarding?

CAN I DESIGN A BETTER PAPER AIRPLANE?
Engineering

PURPOSE: To determine which of three paper airplane designs flies the farthest.

MATERIALS: Paper for planes; tape measure; chart paper.

PROCEDURE:

1. After reading about paper airplanes, choose three designs that you like and that you think will fly well in a straight line. Develop your hypothesis.
2. Make each plane.
3. Set up a test. You will need a long clear space with no wind, and a partner.
4. Make a starting line. You will measure from there forward.
5. From the starting line, throw each plane with the same amount of thrust.
6. Have your partner help you mark down where each plane lands.
7. Throw each plane the same number of times, but at least ten times each.
8. Compile your results. Make a chart. Compare the averages of each plane. Which flew farthest most of the time? Are there any flights that

were exceptional? Tell about them in your results. Try to explain them.

REMEMBER: Tossing each plane with the same thrust may seem hard to do, but it is very important.

RESOURCES: Check your library for books and articles about airplanes, paper airplanes, lift, flight.

THINGS TO THINK ABOUT: Can you compare your designs with those of actual planes? Which would you guess to be fastest? Most efficient? Sturdiest?

WHICH BRIDGE DESIGN IS STRONGEST?

Engineering

PURPOSE: To determine which of three bridge designs supports the most weight.

MATERIALS: Thin pieces of balsa wood; glue; nails; books.

PROCEDURE:

1. Develop your hypothesis.
2. Design three bridge structures to span about 25 cm. You will know more about these after you've done your research.
3. Construct each design out of balsa wood.
4. Set up each bridge across a span. Make sure it will not slip.
5. Add books, such as an encyclopedia, one volume at a time. Continue to add books until the bridge gives way. (See diagram on next page.)
6. Try the same with the other two bridges. Be sure to take photographs before and during the experiment.
7. Which bridge held the most weight?

REMEMBER: You must add the same books in the same order to each bridge to be sure your results are accurate. Weigh the books each bridge held before it collapsed.

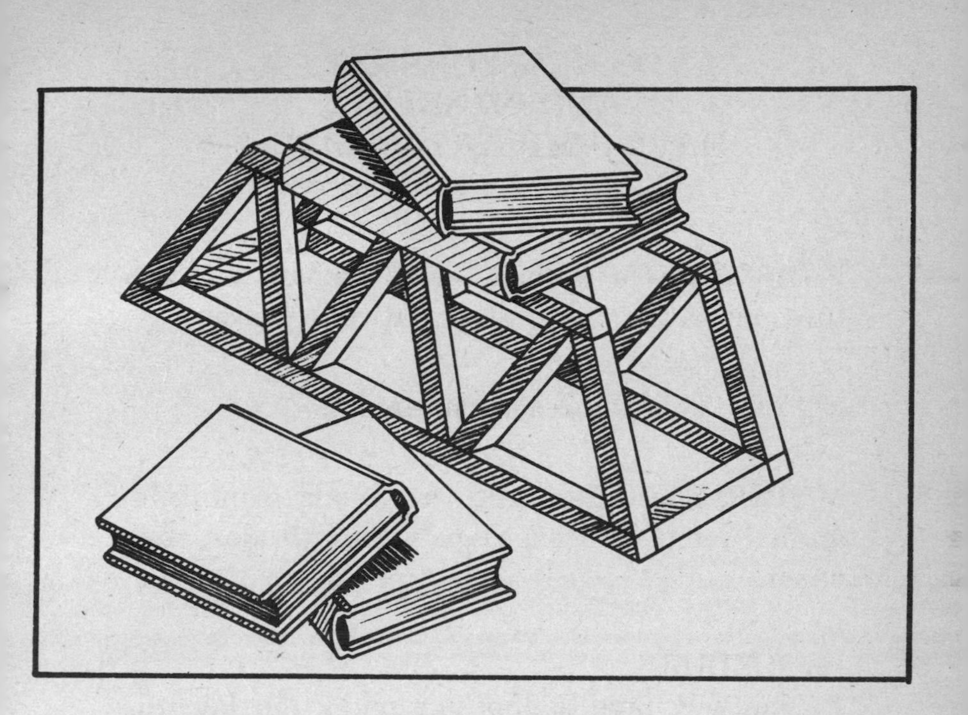

RESOURCES: Check your library for books and articles about bridges, trusses, and bridge building.

THINGS TO THINK ABOUT: Could you design a stronger bridge than famous bridges such as the Golden Gate Bridge or the Brooklyn Bridge?

DO NEW YORK CITY AND ISTANBUL HAVE SIMILAR CLIMATES?

Earth and Space Science

PURPOSE: To determine if annual temperatures are similar at a given latitude worldwide.

NOTE: This is a research project.

MATERIALS: Resource materials containing annual temperatures; maps with latitudes; chart paper.

PROCEDURE:

1. You will have to choose a particular latitude. Then find cities that are along that latitude around the world. Choose several different latitudes. Keep track of each.
2. Make a list of cities for each latitude.
3. Look in another reference book to find the annual temperature ranges for each of the cities on your list. Chart them.
4. Compare your results with each latitude. Do not compare different latitudes with each other.
5. What conclusions can you make?
6. Present your findings in an attractive way. Use maps you've drawn or purchased, make charts, etc.

REMEMBER: A librarian can help you find the resources you will need. She or he cannot do the research for you. You will have to plan on plenty of time researching. The more you research, the better your project will be.

RESOURCES: Check your library for books and articles about temperatures, latitudes, almanacs.

THINGS TO THINK ABOUT: Can you find a city that is the same latitude as yours and compare the two climates?

ARE FOSSILS FOUND ONLY IN SEDIMENTARY ROCKS?

Earth and Space Science

PURPOSE: To determine if fossils are found in types of rocks other than sedimentary ones.

NOTE: This is a research project. You will make use of the library, but you should also check out a science museum in your area, a rock store, or a person known to have a great deal of knowledge about rocks and fossils.

MATERIALS: Research materials; chart paper.

PROCEDURE:

1. Read about fossils and how they are formed. Check for information about types of rocks that frequently contain fossils.
2. Talk to experts. Visit displays. Learn as much as you can about rocks and fossils.
3. What did you find out?
4. Make charts and drawings of your findings. List the types of rock and whether or not they have fossils.

REMEMBER: You will need expert help on this one. The more information you can gather, the better your project will be.

RESOURCES: Check your library for books and articles about fossils, how rocks are formed.

THINGS TO THINK ABOUT: Can you use this knowledge to predict if fossils are likely to be found in the rocks in your area?

WHAT KINDS OF BARRIERS TO WIND EROSION ARE EFFECTIVE?

Earth and Space Science

PURPOSE: To study the effects that different barriers for preventing wind erosion have upon particle size and movement.

MATERIALS: Coarse sand, fine sand, soil, etc.; barriers, such as different screens, foil, tree models; camera and film; fan; flat cardboard box at least 2 feet (60 cm) square and 3 inches (7½ cm) deep.

PROCEDURE:

1. Develop your hypothesis.
2. Set up a scene with about 2 inches (5 cm) of sand or soil in the box.
3. Build and/or add one type of barrier, in the middle of the box.
4. Turn on the fan and let it run for five to ten minutes, depending upon the action you observe.
5. Photograph the changes in your soil or sand. Be sure to notice what happens both in front of and behind the barrier.
6. Repeat steps 2–5 with several different barriers in the wind path.
7. Compare your results.

REMEMBER: Wear goggles to protect your eyes while the sand particles are blowing.

RESOURCES: Check your library for books and articles about the Dust Bowl, wind erosion, landscaping, farming conservation, fencing.

THINGS TO THINK ABOUT: Which type of barrier would last the longest in a field? Which will be most expensive to set up? Which would be least expensive? Which would be least expensive over a span of twenty years?

WHICH WARMS FASTER, THE BEACH OR THE OCEAN?

Earth and Space Science

PURPOSE: To determine which warms faster, sand or water.

MATERIALS: Two identical small containers, such as paper cups; sand; water; thermometers; refrigerator or cool spot.

PROCEDURE:

1. Develop your hypothesis.
2. Fill each container with the same amount of either dry sand or water.
3. Place in a cool area or in a refrigerator overnight.
4. Remove the containers from the refrigerator and set out in the room.
5. Record the temperature of each. They should be the same.
6. Record the temperatures of the sand and of the water every five minutes until they reach room temperature.
7. Make a chart of your results.

REMEMBER: Stick the thermometer into the water or the sand to the same depth in order to control your variables. How can you make sure you are not measuring the moisture in your sand? What kind of sand should you start with?

RESOURCES: Check your library for books and articles about heating and cooling of water and sand.

THINGS TO THINK ABOUT: Could you construct a project testing different colors of sand? Testing wet versus dry sand?

IS THE WATER IN MY AREA HARD OR SOFT?

Environmental Science

PURPOSE: To determine if the water in a given area is hard or soft.

MATERIALS: Ivory soap solution (made from 1 gm soap flakes, scraped from a bar of Ivory soap, mixed with 100 ml distilled water); eye droppers; test tubes; distilled water; tap water.

PROCEDURE:
1. Develop your hypothesis.
2. Add 1 drop of soap solution to 10 ml tap water in a test tube. Shake and see if suds develop.
3. Keep adding 1 drop and then shaking.
4. Record the number of drops it takes for 1 cm of suds to develop in a 10-ml sample of tap water.
5. Repeat the experiment using distilled water.

REMEMBER: The distilled water will be your control. The more samples you can obtain and try, the better your project will be.

RESOURCES: Check your library for books and articles about hard and soft water, and about mineralization.

THINGS TO THINK ABOUT: From your research, you should be able to answer these questions: What makes water hard or soft? How can you predict whether water will be hard or soft?

ARE BIODEGRADABLE PLASTIC GARBAGE BAGS REALLY BIODEGRADABLE?

Environmental Science

PURPOSE: To determine how weather affects deterioration of biodegradable plastic bags.

MATERIALS: Several popular brands of garbage bags, including some that claim to be "biodegradable"; weights, from 100 gm to several kg; scissors; string; plastic bucket; indelible marker; duct tape.

PROCEDURE:

1. Develop your hypothesis.
2. Expose the bags to full sunlight and all weather conditions for about three months. Leave one biodegradable bag and one regular bag unexposed. Your unexposed bags are your controls.
3. Cut each bag into several strips, 5 cm × 30 cm. Label them with an indelible marker.
4. Attach one end of a strip to a board suspended between two chairs. Tape a plastic bucket to the other end. (See diagram on next page.)
5. Add weights to the bucket until the strip breaks. Note the weight.
6. Repeat steps 4 and 5, using several strips from each bag.
7. Compare your results.

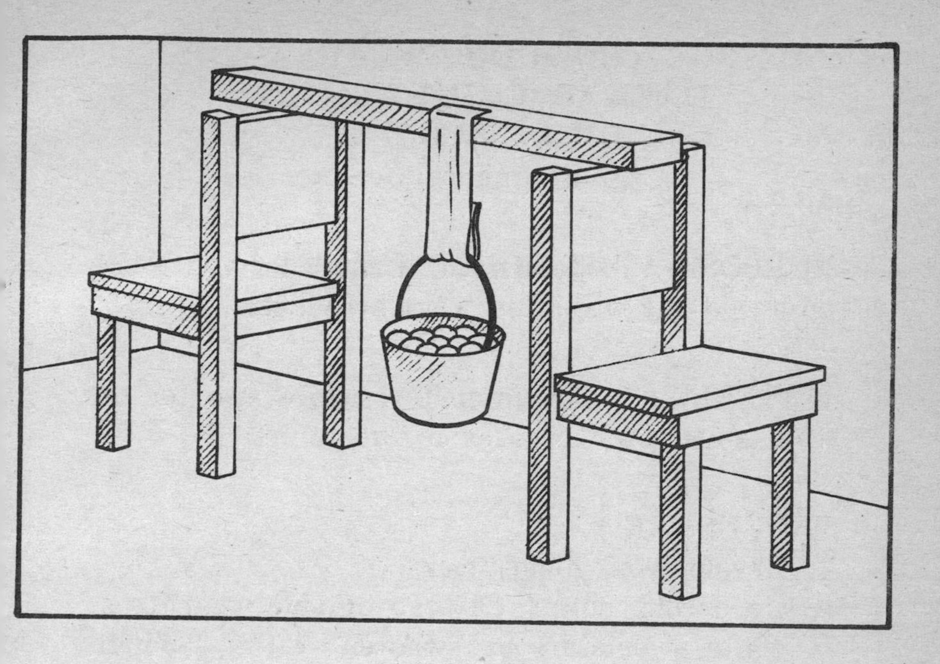

REMEMBER: You will average the weight it took to break each brand's strips. Then compare the averages. You will also want to make notes of how the bags look, feel, etc.

RESOURCES: Check your library for books and articles about biodegradable materials, landfills, pollution.

THINGS TO THINK ABOUT: Which brand would be best if you have many heavy things in your garbage? Which is most ecologically sound? How can you reduce the amount of garbage your family generates?

SHOULD I WEAR DARK OR LIGHT CLOTHES ON A COLD DAY?

Environmental Science

PURPOSE: To determine if the color you wear on a cold day will affect your body heat.

MATERIALS: Thermometer; white sweater or sweatshirt; dark sweater or sweatshirt.

PROCEDURE:

1. Develop your hypothesis.
2. On a cold, sunny day, go outside wearing a white sweatshirt or sweater over a T-shirt. Stand or sit in the sun.
3. After about five minutes, hold the thermometer under your sweatshirt for 30 seconds.
4. Record this temperature.
5. Repeat steps 1–3 wearing a black or dark sweatshirt or sweater over your T-shirt.
6. Compare the temperatures.
7. Repeat this experiment several times and average the temperatures.

REMEMBER: If the sun goes behind a cloud, you will have to start over with your experiment. It is important to control the variables as much as possible. Try to wear sweaters or sweatshirts that are solid colors and that are the same weight. If they do have patterns, wear them inside out.

RESOURCES: Check your library for books and articles about heat loss, insulation, heat, cold.

THINGS TO THINK ABOUT: What colors of clothing make the most sense for winter weather? Summer weather?

DO SNAILS CONTROL ALGAE?

Zoology

PURPOSE: To determine if different types of snails control algae growth.

MATERIALS: Three identical large, clear glass jars; green pond water; two types of snails — an equal amount, by weight, of each type.

PROCEDURE:

1. Develop your hypothesis.
2. Add an equal amount of pond water to each jar.
3. Add one type of snails to one jar. Label it.
4. Add the other type to other jar. Label it.
5. Label one jar NO SNAILS.
6. Check for growth of algae on the sides of the jars after several weeks.

REMEMBER: Keep the jars in the same place. Your research will tell you the best temperature for growing algae.

RESOURCES: Check your library for books and articles about algae, aquarium care, snails, ponds.

THINGS TO THINK ABOUT: Which type of snail would be best to help clean an aquarium? Be sure you check on the compatibility of your snails and fish before you answer that question.

HOW FAST IS AN EARTHWORM?

Zoology

PURPOSE: To determine the average distance an earthworm creeps in one hour.

MATERIALS: Ant farm container (or two sheets of thin Plexiglas with ⅜-inch wooden dowels to form sides); earthworm; moist potting soil; china marker (a wax or grease pencil); tape measure (centimeters); recording sheet.

PROCEDURE:

1. Develop your hypothesis.
2. Fill the ant farm or Plexiglas container with moist potting soil. Do not pack it too tightly. Add the earthworm. Tape down the sides if you used sheets of Plexiglas to keep the soil and earthworm inside.
3. With a China marker, mark on the container the location of the earthworm's head.
4. Check the soil in one hour. Mark the trail of the earthworm. Use the tape measure to measure, in centimeters, the entire trail the earthworm crept. Include any loops or bumps in the trail. (See diagram on next page.)
5. Smooth the soil and repeat the steps. Try to maintain an equal temperature.
6. Keep a project notebook. Record your findings.

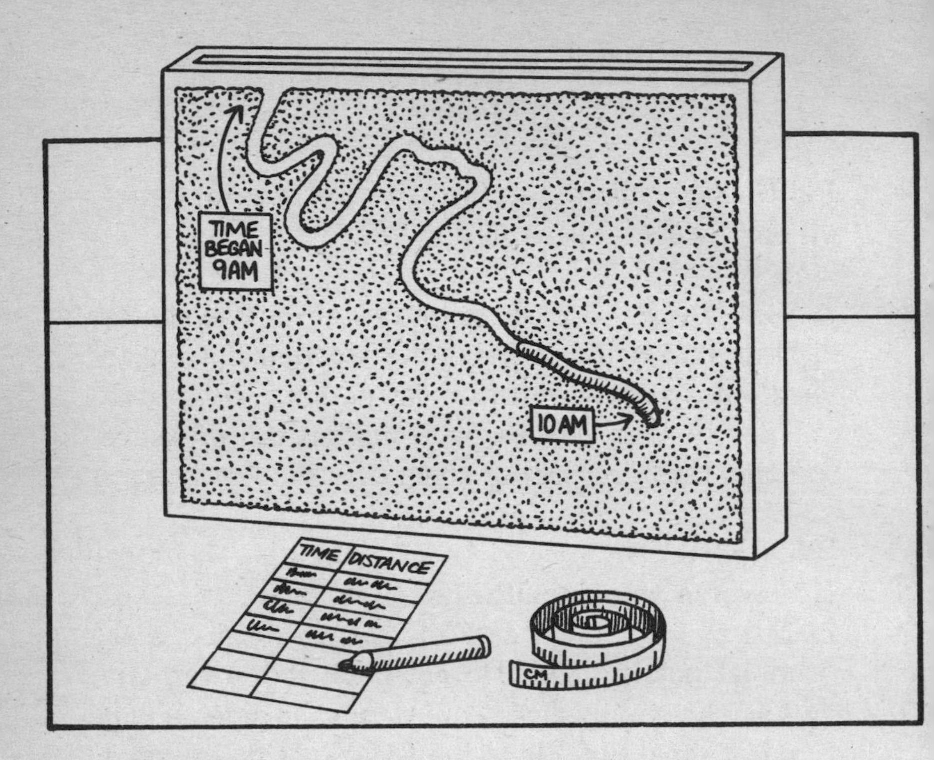

REMEMBER: This is an experiment you must repeat several times. You will *average* the distances. You will make a chart from all the recordings. Repeat the experiment at different times of the day and on different days.

Notes on what the earthworm does when you observe it will be interesting to have in your project notebook. They may lead you to another project.

You can buy earthworms at fishing supply stores any time of year. When you have finished with your earthworm, put it in your garden.

RESOURCES: Check your library for books with information about earthworms, soil, organic gardening.

THINGS TO THINK ABOUT: Why are earthworms good for a garden? How can you make sure they are happy in your garden?

About the Author

Penny Raife Durant is a free-lance writer from Albuquerque, New Mexico, with a master's degree in education from the University of New Mexico. Her ten years in education include teaching in elementary school and teaching and directing a preschool. She is the author of *Make a Splash! Science Activities with Liquids,* and coauthor of *Cinnamon Smoke: Science Activities for Young Children,* and has written for *Parents* and *Working Parents* magazines. She has been a judge for elementary school science fairs, and has helped her two sons with science fair projects.

Paul Mitschler, a consultant to this project, has been a teacher for twenty-seven years with the Topeka Public Schools and Albuquerque Public Schools. He has taught science at every grade level from kindergarten through twelfth grade. He received a bachelor's degree in botany and a master's degree in education from the University of Kansas. He has been a science fair judge at the secondary school level.